ANTI-GENETIX

Anti-GenetiX

The emergence of the anti-GM movement

DERRICK A. PURDUE

Ashgate

Aldershot • Burlington USA • Singapore • Sydney

Published by
Ashgate Publishing Limited
Gower House
Croft Road
Aldershot
Hampshire GU11 3HR
England

Ashgate Publishing Company
131 Main Street
Burlington
Vermont 05401
USA

Ashgate website: http://www.ashgate.com

British Library Cataloguing in Publication Data
Purdue, Derrick A.
Anti-genetiX : the emergence of the anti-GM movement. - (Ashgate studies in environmental policy and practice)
1.Agricultural biotechnology - Social aspects 2.Genetic engineering 3.Transgenic organisms - Patents 4.Environmentalism 5.Social movement
I.Title
363.1'92'5

Library of Congress Control Number: 00-132842

ISBN 0 7546 1216 3

Printed and bound by Athenaeum Press, Ltd.,
Gateshead, Tyne & Wear.

Contents

List of Tables

Preface

Preface

The origins of this book lie in an essay written for my MSc on Ecology and Society in the winter of 1992, testing John Rawls' theory of justice (Rawls, 1973) against the social and environmental implications of the General Agreement on Trade and Tariffs (GATT), for which I read an excellent book by Kevin Watkins on the inequities of the GATT (Watkins, 1992). This project then expanded into a MSc dissertation on seed patenting in the GATT (Purdue, 1993). The emphasis shifted to social movements and it continued to grow into a PhD thesis (Purdue, 1998). Three chapters have devolved into separate publications on social movement theory (Purdue, 1995b), a National Consensus Conference on Biotechnology (Purdue, 1996; 1999) and the neo-tribal identities of local seed savers (Purdue, 2000), leaving the core argument of the thesis in an updated and clarified form as this book. It was prompted by the emergence of two global issues - the patenting of life forms via biotechnology; and the global crisis of natural and agricultural biodiversity. The rationale for this book is to explore the emergence of a social movement in response to these issues by connecting three usually disconnected theoretical literatures on expert systems, global governance and new social movements, which can themselves benefit from being brought together. Thus the book is structured so as to develop these concerns through a case study. Chapter one introduces these empirical issues and theoretical frameworks and discusses the appropriateness of a case study of the anti-GM movement.

Chapter two addresses the dominant discourse of intellectual property rights, as it traces the problematic application of patenting to genes and the consolidation of expert systems. I argue that the dual nature of a patent as a knowledge claim and a property relation binds together an economic drive to commodification and a rationalisation of knowledge into an expert system. Edible seeds may be owned not only as tangible property (food), but also as pure information (genetics). Intellectual property is both an expert system itself and the ownership of expertise in general. Intellectual property and expert knowledge are self-referential texts, but lay claims to universal applicability and priority over other knowledges. The new knowledge embodied in patents is new to that expert system and has to take the form of the expert discourse itself, but it is not necessarily new to other knowledge systems. The property rights and expert knowledge are mutually reinforcing. The major criticism of gene patenting is that it converts commonly held resources into private property, forcibly separating these resources from their previous owners. While patenting genes has been the subject of strenuous opposition, another form of intellectual property, Plant Breeders' Rights (PBRs) had already been developed specifically to apply to plant varieties, and therefore seeds. The invasion of patenting into the territory of living organisms has been accompanied by a tightening of PBRs to approximate patents ever more closely; indeed the two are tied together in the TRIPs agreement.

Chapter three explores shifts in global governance, focusing on seed patenting in the TRIPs Agreement. The institutionalisation of neo-liberal hegemony with the formation of the World Trade Organization (WTO) during the Uruguay Round of the GATT and Intellectual Property Rights extended to life forms in the TRIPs Agreement. TRIPs is a nodal point in the construction of a global regime in property rights over seeds, which includes other new institutions such as the Commission on Biodiversity (CBD) and older ones such as the Food and Agriculture Organization (FAO).

In Chapter four I map the counter-expert networks in the Non-Governmental Organizations (NGOs) that campaign against patenting life and related issues in the UK and analyse the ways activists frame the issues they are concerned with. The network has no definite centre, consists of individuals with varied relations with their organizations, rather than a network of organizations as such. Trust plays a key role in maintaining this fragile form of organization. The activists have a dynamic role in relation to national and global institutions and to the public. NGOs in the first place alert a wider public (and media) of the impending changes and shape the inchoate anxieties about biotechnology into frames which link safety, equity and integrity of organisms. Social movements consist of three parts, visible mobilisation, submerged cultural networks and counter-expert networks that engage directly with authoritative expert systems. The current cycle of protest against GM food in the UK evolved from a core of intensely active counter-experts.

Chapters five and six explore the role of social movements in global civil society through case studies of a major NGO gathering in Bern and the NGO mobilisation at the 4th Technical Conference of the FAO in Leipzig. The anti-GM movement's involvement with the Leipzig process consisted of two principal phases. The first phase consisted of a consensus-building NGO meeting where differences were voiced and discussed so that representatives of large constituencies in the South and counter-experts from the North could mesh their political and knowledge claims. The second phase consisted of NGOs working together in co-ordinated lobbying, fringe activities and press work at the intergovernmental conference. The NGOs occupied a position along a spectrum of marginal inclusion in the layering of political power through the FAO. They displayed a 'prophetic' role as leaders of global civil society, taking command of the theoretical and moral heights left open by the empty technical diplomatic discourse of the international organization. The outlines of an emerging global movement came into view. These networks have widened and fed into the spectacular mobilization against the WTO in Seattle in December 1999.

Chapter seven returns to the core theoretical concerns of the book. Global hegemonic projects play through the institutions of global governance. Such projects take the form of discourse-coalitions mobilising various actors and expert systems around key story lines that promote special interests as global necessities. Global social movements are emerging that encounter hegemonic projects as their adversaries. Relatively small networks of counter-experts, such as those that formed the early leadership of the anti-GM movement, play an important role in global politics, acting to change international regimes and consolidate global civil society.

Acknowledgements

I would like to thank the Centre for Social and Economic Research at UWE for providing me with a bursary for the PhD research on which this book is based. I enjoyed the support of an exemplary supervisory team. Thanks to my Director of Studies, Professor Peter Glasner, and my supervisors, Dr Ian Welsh and Peter Jowers for their support during my thesis and suggestions for its transformation into a book.

Thanks are also due to Professor Murray Stewart and the Cities Research Centre at UWE for their support during the preparation of this book, as well as Christine Taylor for producing camera ready copy. Professor Michael Waller provided editorial comments on an earlier version of chapter three that has been published as 'Hegemonic Trips: World Trade, Intellectual Property and Biodiversity', *Environmental Politics*, 1995, 4 (1): 88-107.

The book would not have been possible without the campaigners who were so generous with their time and information allowing me to interview them and come to their meetings. Patrick Mulvany deserves a particular mention for keeping me in touch with the developing NGO networks and making my trip to Leipzig possible as well as being an invaluable source of information. Thanks also to Robin Jenkins who was a very useful informant in Bern and Julie Hill who invited me to a Green Alliance seminar.

My partner, Pippa Adamson, not only proof read the final draft, but gave me immeasurable support throughout while living with the book for the duration, arguably the hardest job of all. Finally, my sons, Innes and Callum, continue to inspire me with their creativity and keep my feet on the ground.

List of Abbreviations

ABC	Agro-Biodiversity Coalition
ACRE	Advisory Committee on Releases to the Environment
ANT	Actor Network Theory
A-SEED	Action for Sustainability, Equality, Environment and Development
BUAV	British Union Against Vivisection
BUKO	Bindeskongres entwicklungspolitischer Aktiongruppen
CBD	The Convention on Biological Diversity
CEAT	The Co-ordination Europeenne des Amis de la Terre
CGIAR	Consultative Group on International Agricultural Research
DNA	Deoxyribose Nucleic Acid
EPO	European Patent Office
FAO	The Food and Agriculture Organization
FIELD	Foundation for International Environmental Law and Development
GATT	The General Agreement on Trade and Tariffs
GENET	Genetic European Network
GIG	Genetics Interest Group
GMO	Genetically Modified Organism
GRAIN	Genetic Resources International
HDRA	Henry Doubleday Research Association
HGDP	Human Genome Diversity Project
IARCs	International Agricultural Research Centres
IBA	Industrial Biotechnology Association
ICDA	International Coalition of Development Agencies
IPRs	Intellectual Property Rights
IR	International Relations
IMF	International Monetary Fund
IT	Intermediate Technology Development Group
IWC	International Whaling Commission
LETS	Local Exchange Trading Systems
NAFTA	North American Free Trade Agreement
NGOs	Non-Governmental Organizations

PBRs	Plant Breeders' Rights
PGR	Plant Genetic Resources
PGRFA	Plant Genetic Resources for Food and Agriculture
PGS	Plant Genetic Systems
PMA	Pharmaceutical Manufacturers Association
RAFI	Rural Advancement Foundation International
rBGH	Recombinant Bovine Growth Hormone
rBST	Recombinant Bovine Somatropine (rBGH).
rDNA	Recombinant Deoxyribose Nucleic Acid
SAFE	Sustainable Agriculture Food and Environment
SAGB	Senior Advisory Group on Biotechnology
SWISSAID	Swiss Aid Agency
TNC	Transnational Corporation
TRIPs	The Agreement on Trade-Related Aspects of Intellectual Property Rights
UNCED	United Nations Commission for Environment and Development
UNESCO	United Nations Education, Science and Culture Organization
UPOV	The Union for the Protection of New Plant Varieties
WIPO	World Intellectual Property Organization
WRI	World Resources Institute
WTO	The World Trade Organization
WWF	World Wide Fund for Nature

1 Genetic Patenting: Knowledge, Global Governance and the Anti-GM Movement

The Issues: GM Food and Seeds

Biotechnology and Gene Patenting

Since the mid-1970s a series of techniques have been developed that allow DNA sequences, or genes, to be isolated and transferred from one organism to another. These techniques, known collectively as genetic engineering or biotechnology, can breakdown barriers between species that are not naturally part of the same breeding population. Biotechnology produces transgenic or Genetically Modified Organisms (GMOs) such as sheep which contain genes drawn from humans; or potatoes that contain genes from moths, a lettuce that has a gene from a fish (Baumann et al, 1996). There is some debate as to the effect of a gene when transferred into a quite different organism (Kolleck, 1994), especially given that genes can interact with each other and the outcomes can be influenced by environmental issues. Biotechnology is a new technology for which ambitious claims are made; yet there are high levels of uncertainty accompanied by increasing public anxiety.

GM (genetically modified) food became the subject of public debate in the late 1990s as it began to arrive in supermarkets. Much of this discussion focused on potential health risks associated with the uncertainty inherent in the technology. In the UK, BSE and to a lesser extent other food crises, have reduced public trust in the food industry and government regulation. In this context GM food has often taken the shape of a consumer protection issue. Yet biotechnology has potentially far-reaching impacts for food producers as well, particularly in the Third World, through what might seem like the back door, namely patents. Patents are the strongest form of Intellectual Property Rights (IPRs) through which technological ideas can be owned and rented ou to others. Laying claim to new patents is a potent economic driver in this whole new field of scientific and technological endeavour, which may well hold the key to the genetic base of agriculture in the twenty first century. The majority of GMOs produced, patented and released are plants. From this angle, patenting seeds puts corporate profit before sustainable agriculture or social justice for Third World farmers.

Advocates of biotechnology claim it can produce the super-crops that will feed the world, by introducing new traits to food crops, such as resistance to drought, frost or pests. In fact, the development of genetically engineered crops has been driven by their patentability and the profits these patents are hoped to generate, with little concern for social justice central to any viable solution to food shortages. A common strategy has been for a company to patent new varieties that are resistant to the herbicides already patented by the same company. Monsanto, for example, have produced a matched pair with their herbicide Roundup, and Roundup-resistant soya beans. Rather than crucial innovation, much of this corporate effort is directed to locking farmers into an industrial food cycle where they pay rent on genetic material that they no longer control. Patenting also allows the privatization of presently available crops with minor genetic change, as has been done with the spice turmeric or basmati rice, both widely used in Indian cooking, and parts of the Neem tree long used as toothpaste in India. Patents are pending on potatoes and chickpeas amongst many others.

Patents have in the past been hotly contended as a way of controlling the spread of previous waves of technology, from Britain to America and later from America to Japan. Now the fault line runs between First World and Third World, with less than 1% of patents held by Third World nationals (Ektowitz and Webster, 1995). Patents are popularly associated with the invention of obscure machines that do something nobody can yet see any use for or at best hopeful get rich schemes. Patents did indeed traditionally apply to machines, specifically machines that could be demonstrated to be new and to be beyond the scope of a skilled craft person to produce without prior knowledge. The patent itself is actually a description of how to make something NEW and useful. It explains a new process and is therefore a form of knowledge. A patent discloses, or makes public, a new parcel of knowledge, but it also declares the ownership of that parcel of knowledge. No one else can use the knowledge without permission and usually payment.

The moral argument for patenting is that your invention is a product of your mind and therefore you have an inalienable right to own it and any profits made from it. The economic argument which is now widely used is that the large companies that own most patents have spent a lot of money on research and deserve to get their money back in monopoly rights and/or rents on their inventions. Patents have two interesting limits. In theory, you cannot patent what is already publicly known. Nor can natural processes be patented, since scientists may discover them, but engineers cannot invent them. With the development of genetic engineering, patent law has been extended to cover living organisms and their genes, which has put both these limits to the test.

Genes are in one sense physical objects that may be eaten, for example as one dares to eat a peach. But in another they are virtual objects consisting primarily of information, codes for triggering physical effects in an organism, through the manufacture of protein. Similarly, seeds are natural objects which reproduce independently of human thought or intervention, yet humans have for centuries been shaping the genetic composition of the seeds they use to grow food with interesting colours, flavours, textures or the ability to survive frost or drought.

The landmark case of gene patenting came in the USA in 1980 [Diamond v. Chakrabarty] in which a bacterium was successfully patented on the grounds that it had a novel gene inserted. Legal battles have ensued over whether genetically modified animals, micro-organisms and plants are best seen as naturally occurring organisms, part of nature which may be studied, but not invented, or technologies bearing the imprint of human intentions. The borderline between scientific study and technological innovation is very fuzzy. European legislators have, for many years, been divided over whether human genes are body parts or strings of information. Body parts like your hand or your liver clearly cannot belong to a company under a patent. Strings of information can be used to produce a new invention, such as a mouse prone to breast cancer, patented by Du Pont and Harvard University. Just as it has long been difficult to separate science from technology, so it is increasingly difficult to separate technoscience from Intellectual Property Rights, genetics from patents.

Most countries in the South had until recently no intellectual property law covering seeds, either in the form of PBRs or patents. The seed trade in the South is divided into three main parts. (1) A global trade in F1 Hybrids, rapidly concentrating in the hands of very few transnational corporations (TNCs) usually agro-chemical companies with strong pharmaceutical interests, and increasingly, biotechnology subsidiaries. (2) Domestic seed companies trading in open pollinated seeds. (3) Local farmers saving and exchanging outside of any control by seed companies. It has been estimated that over 80% of seeds in Southern countries stay outside the official trade (van Wyk, 1995). As such, they form a potentially lucrative source of new royalties for seed companies.

Biodiversity and Sustainable Argriculture

The concept of 'Biodiversity' was invented in the mid-1980s as a way of articulating a sense of crisis about the catastrophically rapid loss of wild species of plants and animals, and indeed whole eco-systems (Wilson, 1988). Tropical rain forests became emblematic of this aspect of global ecological crisis. Part of the argument mounted to defend endangered species was that poor people in the South (particularly indigenous people) utilise the wide variety of natural life forms and eco-systems available to them. Biodiversity became enshrined in international law at the Earth Summit in Rio de Janeiro in June 1992, when the Biodiversity Convention was signed by an unprecedented number of heads of state. Critics suggested that sudden eagerness to protect diversity, increasingly thought of as genetic diversity, coincided with the development of the biotechnology industry in Northern countries which could potentially turn rain forest genes into profits (Shiva, 1993).

From the late 1970s, NGOs began to perceive a similar crisis developing within agriculture, where domesticated varieties, species and whole farming systems were disappearing. Again it was argued that the poor in the South were principal losers, as this diversity was once again a key set of resources in maintaining their livelihoods. Agricultural biodiversity may be defined as the global stock of plant genetic resources, mainly consisting of the varieties of crops cultivated in the Third World and their wild relatives, such as the thousands of potato species and varieties grown in the Andes. This biodiversity is vital to maintaining a varied agriculture and secure food supply for poor people.

The emergent anti-GM movement appropriated the frame of biodiversity to insist that agricultural biodiversity is, if anything, more important than 'natural' biodiversity. The value of biodiversity in crops is that genetic variation allows adaptation to changing conditions, such as weather, pests and weeds, as well as changing human food needs and desires - growing season, taste, texture, suitability for different cooking styles. In Britain for example, varieties of apple have been bred to fruit consecutively through most of the year.

There are five principal causes of crop biodiversity loss in the UK.

(a) The growing patterns of industrialized agriculture tend to homogeneity.

(b) The control of the seed industry has been concentrated in the hands of a few transnational agro-chemical and pharmaceutical companies. The number of seed suppliers has declined rapidly as the seed trade has globalized. The future of plant breeding is seen as depending on biotechnology. Transnational agro-chemical companies have first captured biotechnology companies and now the seed trade. Pioneer Hi-Bred is the only one of the top ten global seed companies that is not owned by an agro-chemical company. While the agro-chemical companies control the cutting edge technology, seed companies have the distribution networks and vital genetic resources, including the inbred 'elite' varieties that provide the parentage of the successful commercial hybrids (Kloppenburg, 1988a). By 1996, there were only 6 major seed companies left in the UK not owned by transnationals (King, 1996). Austria lost all its seed companies within a year of joining the EU. The only source of local seeds was an NGO (Cherfas *et al*, 1996).

(c) The EU Common Catalogue specifies the tested and registered varieties that may be legally sold in the EU. To be included on the list, a variety must be tested for distinctiveness, uniformity and stability. The owner of the seed must pay for the testing and an ongoing registration fee in exchange for a Plant Breeder Right over variety, which grants the owner a monopoly of the market and hence royalty payments. Rare varieties are thus too expensive too maintain due to the small volume of sales.

(d) Seed saving is increasingly difficult for farmers due to the commercial use of F1 Hybrids that don't breed true.

(e) Ready access to packaged seed through garden centres and catalogues discourages gardeners from saving their own seed. Seed saving has been eclipsed by the commercialization of gardening in general with the expansion of garden centres, reorganising gardening around consumer pleasures such as choosing excitingly packaged seeds, rather than saving and re-planting. Hence the de-skilling of agriculture and gardening associated with dependence on expert systems, runs parallel to that experienced in other areas of everyday life, such as cooking. Thus many of the seed savers I have interviewed have no familial connection with seed saving, but have come to it for ideological reasons as part of a global debate on biodiversity (Purdue, 2000).

Erosion of agricultural biodiversity is not, however, simply a matter of heavy-handed regulation. In the USA, where 'Farmers Varieties' and 'Heritage Varieties' are marketable, 90% of fruit and vegetable varieties available at the turn of the century have still been lost (Fowler and Mooney, 1990). The vertical integration of the seed and agro-chemical industries allows a company, such as Monsanto, to use biotechnology to develop and patent a genetically engineered variety of soya beans, which is resistant to its own in-house herbicide, Roundup. This soya bean went into the UK food chain at Christmas 1996. The monoculturalist tendency of industrialized agriculture combines with new technology, economic globalization and regulation of Intellectual Property Rights (IPRs) to produce a galloping biodiversity crisis in agriculture.

At the global level the causes of agricultural biodiversity loss are similar, with the rapid increase in industrialized agriculture utilising F1 Hybrids in monocultures. However, patenting threatens to add a new dimension. Patenting of seeds became possible with the application of genetic engineering techniques to plants. A patent means that not only must a royalty be paid on purchase of the original seed, but that any seed that a farmer grows and then re-plants or sells, would infringe the patent. Many of the patents currently held are for varieties that are resistant to a herbicide produced by the same company as the seeds. Critics believe that the move to patenting will tie farmers more tightly into a chemically driven farming system, which is not in the long term interests of producers or consumers. Small farmers in the South will lose genetic diversity of their crops, and with biodiversity loss, they will lose their ability to reduce the risks posed changing conditions.

Theoretical Frameworks

The Commons, Citizen Science and Expert Systems

One way of seeing the changes wrought by patenting is by using the analogy of the village commons on which British people were able to graze their animals. No one individual owned exclusive rights to the land, but many had access to it. These commons were later fenced off by powerful landowners to the detriment of local users (Thompson, 1991).

While seeds have long been bought, sold and bartered, the genetic material in seeds traditionally formed a commons in that once acquired and planted, seed could be used to grow new seed for replanting. Farmers could use the seed to select for special characteristics and crossbreed with other varieties. These innovations would become part of the stock of common genetic resources embodied in the seeds, available for further innovation by subsequent farmers. Patenting stops all that. If you buy a patented seed, not only are you required to pay royalties to the patentee, but you infringe the patent if you attempt to use the seed you grow from it to replant the next year's crop, or to breed new varieties.

The tradition of public science represents a second form of commons. When scientists publish their findings in the form of a scientific paper, the academic reputation they gain from their work is separated from the knowledge they produce. The content of the paper becomes common property which others may use for research or to develop new products. Publication as a patent on the other hand encloses this knowledge as private (usually corporate) property, which cannot be used without agreement and royalty payments. This may sound like a quick route to wealth for university researchers. In odd cases this may be true. However, many patents barely pay for the patent fee. In the UK only 25% of patents are still in force after 5 years, 5% run the full 20 years (Newton, 1997). Nevertheless, considerable pressure has built up for public sector scientists to patent their work, and holding patents has become a criterion for judging their merit. Biotechnology in the USA has been typified by professors setting up private companies to cash in on the work done (often by their graduate students) in public institutions (Kloppenburg, 1988a). In collaborations between private and public sectors corporate investment is usually tied to holding all patent rights and so also control of publishing. Zeneca, for example, have made it quite clear that without patents they will not collaborate (Purdue, 1996; 1999). Clearly, such arrangements disrupt the free flow of information and scientific collaboration further eroding the scientific commons.

Knowledge is not solely the province of professional scientists. Lay citizens of all sorts produce and use local forms of knowledge, which has been called citizen science (Irwin, 1995). Farmers and gardeners are often lay geneticists in as far as they select and cross varieties to produce crops suitable to local conditions. These forms of citizen science are crucial to food production around the world and central to any attempt to support poor countries feeding themselves, yet are not recognized in patent law as scientific knowledge. Patents draw a hard line dividing corporate scientists from knowledgeable lay citizens around the globe. With patents in force, farmers will lose their legal rights to pursue their citizen science or to use their knowledge commercially. Those who have developed useful seeds by traditional means will not be financially rewarded. The companies that use molecular biology to make the final genetic move that is patentable will hold monopoly rights to the seeds. First World companies dealing in seed, agro-chemicals and biotechnology, are using patents to enclose the Third World's common genetic resources and common citizen science.

This kind of disembedding of local knowledge and its transfer to a global discourse overseen by professionals highlights the power that falls to experts (Beck, 1992), or more accurately expert systems (Giddens, 1990), in contemporary society. The lay public is required to trust expert systems to regulate risks and re-embed the disembedded expert knowledge into their daily lives. However, expert systems colonise citizen science produced outside its professional and conceptual borders. That is, they simultaneously capture such knowledge and undermine its legitimacy. Yet key social movement activists act as counter-experts (Jamison and Eyerman, 1991), who challenge the social construction of a binary division between expert systems and a deskilled lay public.

Global Governance

While the legal debates continued over whether a genetically modified seed was an invented technology, which could be patented, or a self-reproducing organism which could not, the GATT Uruguay Round was signed in Geneva in December, 1993, creating the World Trade Organization (WTO). The WTO included a sub-agreement on Trade Related Intellectual Property (TRIPs), which required all member states to allow for the patenting of seeds and genes, or produce their own sui generis system of equivalent effectiveness. Significant political pressure was brought to bear in order to close the sui generis loophole by defining ‘equivalent effectiveness’ in terms of Plant Breeders’ Rights as defined by another international organization, the Union for the Protection of Plant Varieties (UPOV). UPOV, a relatively small group of rich countries, had in 1991, produced a definition of PBRs almost indistinguishable from patents, effectively forcing Third World countries to accept global patent law.

Within the UN system, the Food and Agriculture Organization (FAO) had long been concerned with genetic diversity in seeds. The FAO called its 4th Technical Conference in Leipzig in June 1996 in order to consider a Report of the State of the World's Plant Genetic Resources and to formulate a Global Plan of Action to conserve these resources. It was intended that an Undertaking on Farmers' Rights in respect to seeds would also be agreed by the FAO and sent to the Convention on Biological Diversity to be included as a binding protocol. A web of international organizations are engaged in a series of interlocking agreements that are defining the terms on which seeds are grown, eaten and replanted in the near future. This apparently esoteric set of agreements will have complex, but tangible, potentially devastating effects on the food security of much of the world's human population.

The concept of governance has been used to indicate a steady shift from government to greater participation of first the private sector, and then civil society in the policy process, across a wide range of policy areas. Governance may be local (Stoker, 1998), national (Jessop, 1995) or global (Giddens, 1998). Yet global governance presents peculiar problems. International relations as a discipline has always been absorbed by the problem that international society exists without a world government (Shaw, 1993). International agreements and institutions that have emerged from states have co-operating to deal with international or global problems have come to be known as international regimes. Different theories suggest that in regimes (1) each state acts as a self interested actor, or (2) the regime is dominated by a hegemonic power, or (3) all member states enter a partnership for collective gain (Young, 1989). An interlocking network of international regimes - the World Trade Organization (WTO), the Commission on Biological Diversity (CBD), UPOV and the Food and Agriculture Organization (FAO) – now regulate patenting and biodiversity globally. This network constitutes a new form of global governance. While governance may seem more open to influence than an institutionally unified government, the same powerful actors and discourses play through all the separate fora. The positivist model on global hegemony available in International Relations theory depends too heavily of a single state acting as undisputed hegemon. A post-structuralist model of incomplete hegemony (Laclau, 1990) and hegemonic projects (Hall, 1988), developed at a national level, needs to be extended to global hegemonic projects. Similarly, regulatory regimes are not discreet objects, but texts referring to a web of other texts in a global intertext (Der Derian 1992). While the USA may not be an undisputed global hegemon, a hegemonic discourse consisting of the neo-liberal 'governmentalities' (Thrift, 1999) that have restructured economies and reshaped attitudes to property rights, including intellectual property rights, runs through global governance. The economic interests of the biotechnology industry, particularly the agro-chemical transnationals, framed in the expert discourse of patent law, are woven into the project of making IPRs over genes and seeds globally uniform.

This hegemonic project is a discourse coalition (Hajer, 1995) of a complex articulation of scientific, biological, legal and economic expert discourses, linking social actors such as biotechnology companies, lobby groups, agro-chemical transnational corporations, seed companies and national governments. The key story line of this coalition is that technology, social progress and corporate profits are indissolubly linked. It draws on a wider 'Promethian discourse' (Dryzek, 1997) that claims that all environmental problems are soluble through technological ingenuity, precisely of the sort that patents are supposed to reward. Opening up this project to public scrutiny is a key role, played by the anti-GM movement.

Social Movements

New social movements develop new collective identities (Melucci, 1989; 1996) and struggle for the acceptance of their identities as participants in the struggle and for tangible gains in terms of the issues that concern them (Della Porta and Diani, 1999). Social movements consist of loosely connected social networks, that embody a collective identity; they discursively frame the issues that concern them as a conflict with one or more adversaries; and they have to develop a repertoire of collective action to engage with their adversaries.

Social movement networks are the 'experimental laboratories' in which new collective movement identities and meanings are explored (Melucci, 1988). These networks are also channels for the circulation of practical resources and meaning; they 'facilitate the organization of both political and cultural conflicts'; and 'may consist of relations between individuals, movement organizations, or both' (Diani, 1996: 8). The anti-GM networks explored in this book were a mixture of individuals and groups which organized politically and culturally through flows of meaning; in particular the definitions of what they were doing and what was at stake, and flows of resources including seeds and genes themselves. These were largely leadership networks consisting of counter-experts. Uncertain, but dynamic, relationships were developing between these counter-experts and wider social movement constituencies amongst Southern peasants and Northern organic farmers and gardeners, as well as scientific experts and governmental delegates, and more recently consumers and environmental protestors. It was significant that these constituencies were physically distant from each other, but part of the same global movement.

A social movement produces a discourse that articulates value judgements about social issues into an 'injustice frame' (Gamson, 1995). The articulation of this injustice opens up a conflict against a dominant adversary and shapes the identity of the movement (Touraine, 1981; 1995; Melucci, 1989, 1995; Jamison and Eyerman, 1991) in an 'adversarial identity frame' (Gamson, 1995). The nature of this adversary has always remained unsatisfactory. Neither Touraine's earlier formulation of technocracy (Touraine, 1981), nor his later formulation of rationalization (Touraine, 1995), have resolved the ambivalence as to whether this dominance consisted of a dominant movement or a technocratic system. Nor have subsequent authors fared any better (Melucci 1989; Jamison and Eyerman, 1991). The dominant adversaries that social movements confront frequently have systemic, discursive and movement characteristics and are best described as hegemonic rather than purely structural forms of power. Global hegemonic projects involve the construction and extension of expert systems, in which particular interests, or cultural constructions, are embodied as universal epistemological and ethical principles. Hegemony cannot be simply defined as an objective social fact. It requires a social movement point of view to make visible the hegemonic power embodied in the business-as-usual operation of technocratic organizations.

In addition to building networks of solidarity and articulating a conflict, a social movement must also develop a sense of its agency (Gamson, 1995) and adopt ways of communicating with its adversary through a repertoire of collective action (Tarrow, 1994). The repertoire developed by social movements since the French Revolution is 'modular', with new modules added in each 'cycle of protest' (Tarrow, 1994). For Touraine and Melucci social movements articulate a conflict that is antagonistic to the dominant social system (Touraine, 1981; 1995) and break the limit of the system (Melucci 1989; 1996). 'Breaking the limit of a system' is often operationalized in empirical terms as the use of disruptive action repertoires (Tarrow, 1994) frequently linked to representation of excluded social groups (Useem and Zald, 1987). From this sort of perspective, there can be no antagonistic social movement without mass mobilization. These settled expectations of what constitutes social movements: mass mobilizations, protest cycles, disruptive action repertoires have already been undermined by work on latent cultural networks (Melucci, 1989). Counter-experts, as movement leaders, frame the concerns of the latent networks and represent these networks as a constituency, by engaging with institutionalized systems.

The collective identity of a social movement cannot be taken for granted, rather it is the outcome of collective action (Melucci, 1989). Complicated processes of negotiation continue between different networks and movement

leaders, with the alignment of frames and choices of action, or at least some sort of accommodation. The pattern of connections within a movement shifts over time as networks expand, fragment or whither, discursive frames diffuse or succeed each other and the action repertoire changes.

The leadership of the anti-GM movement in the mid-1990s depended on a small network of individuals within environmental Non-Governmental Organizations (NGOs) who had developed counter-expertise combining a situated knowledge of seed science and law with impressive political skills. I have chosen to refer to the environmental organizations as NGOs, because that is what they called themselves. In so doing I have avoided the sociological term 'Social Movement Organization', or 'public interest group' either of which are preferable to the more theoretically suspect term 'pressure group'. NGOs are widely recognized as a rapidly growing, increasingly important, though poorly theorised feature of international relations (Princen and Finger, 1994).

Furthermore, the binary opposition between political insiders who participate and outsiders who protest should be avoided. The NGOs were explicitly political, but neither absorbed nor entirely excluded from global institutions. As counter-experts, they provided a rival framing of the ethics, identities and agencies involved in the struggle over the control of gene pools, agricultural practices and living organisms more generally. They have, with varying degrees of success, challenged the hegemonic project of global patenting both in the negotiation of regimes within international law and by contributing to the development of a global civil society. This emerging global civil society is still a fragmented arena, where a range of global public interests and opinions are represented and indeed constructed. These newly articulated global perspectives are often Southern in origin, and in antagonistic conflict with the biotechnology industry's claim to global IPRs. Counter-expert networks constitute a key element in the emergence of a global anti-GM movement. Practical seed savers operate under very different conditions in North and South, but in both cases the definition of their work as an engagement with the hegemonic project of biotechnology and patenting is dependent on the framing work of counter-experts.

Social movement theory has been developed principally to explain the 'new' or 'left-libertarian family of social movements' that emerged in the late sixties and early seventies. In spite of the complexity of these theories they do not deal with movements that engage with global governance. Nor do they explain the possibilities of social movements intimately connected to knowledge of new technological complexities. Just as social movements theory is weak on engagement with global politics, so theories of International Relations and expert systems underplay the role of social movements and the counter-experts who are central to them. Social movement counter-experts play a creative role in global governance, making some impact on international regimes, as well as developing global civil society.

Research Approach: A Multi-Method Case Study of the Anti-GM Movement

A case study may test theory or develop new theory through the exploration of a particular example. Social movement theory has largely been developed out of case studies of national movements. The anti-GM movement varied from what the existing literature on social movements might lead one to expect a social movement to look like on a number of counts, which made it a theoretically interesting case study. (a) It was explicitly global in its networks, framing of issues and in its action. NGOs often dealt less with organs of the nation state than with the evolving mechanisms of global governance - representing a significant route of development for social movements in the next decade. (b) The anti-GM movement confronted a knowledge-intensive, cutting edge technology and was therefore obliged to act as counter-experts. This is indicative of wider tendencies in contemporary society, where expert systems have come to play an increasingly important role in everyday life. (c) The anti-GM movement has re-articulated old issues of distributive justice in new ways, as intellectual property has begun to replace material property as a key to controlling social resources. (d) The anti-GM movement turned taken-as-given areas of social life, such as food and sexual re-production, into problematic areas of new dangers. (e) The movement re-worked the symbolism of local distinctiveness in a global context. (f) Social movements typically consist of several distinct social groups, but the anti-GM movement is unusual in that these groups, such as peasants, counter-experts, organic seed enthusiasts, occur in geographically distinct areas of the globe. Overall, the anti-GM movement both resembled and differed from other contemporary social movements. However, it was not an isolated empirical case of social movement intervention in global politics, as the growth of the 'NGO phenomenon' (Princen and Finger, 1994) and the large mobilizations at the Rio Earth Summit and the Beijing Women's Conference attest. The anti-whaling movement had a similar focus on a global institution (Stoett, 1993) and therefore bore some resemblances to the anti-GM movement. As a case study the anti-GM movement can, therefore, illuminate some of the possible new trajectories in social movements and global politics.

Case studies frequently employ a multi-method approach in order to gain multiple perspectives on the material. These methods have to be chosen according to the nature of the case study undertaken (Rose, 1991). My research methods evolved with the fieldwork. Semi-structured interviews with NGOs

produced a snowball sample from which a core set of key NGO actors, who were cited as important by their peers, began to appear. At this point my interest shifted to participation observation of the process. While I did not attempt to conceal my identity as a social researcher, I did not always openly negotiate an overt research role with all those present at the outset. More typically I would negotiate to attend a meeting with a gatekeeper. In larger meetings it was not feasible to gain the approval of all participants, but then it was usually acknowledged that press were present.

In analysing data a mixture of methods proved useful, including elements of ethnography, qualitative network analysis and discourse analysis. Hence I have identified the key actors and nodal points in the networks and explored the roles of individuals and NGOs and the embodiment of networks in events or meetings. A second layer of analysis concerns discursive frames or story lines and their relation to the construction of a collective movement identity. This includes the framing of issues and identities as well as more affective ways of embodying collectivity.

Framing and networking appear to be two contradictory analytic metaphors: networks imply actors and connections, while framing implies limits and unexamined decisions (Constantinou, 1994). Yet, a network connects around a shared story-line, which focuses attention on key concerns and frames the terms in which these issues will be thought. So for NGOs, framing the issues to be addressed and networking with the individuals deemed relevant occur simultaneously. For an NGO to identify an issue as worth working on involved developing a story-line or discursive frame that could link together relevant issues and simultaneously close off those issues they felt to be unrelated or impractical to deal with. Issues and networks also interacted with shaping an action repertoire: the types of activity an NGO chose to pursue and whether they pitched their work at the local, national, European or global level. A whole set of subsidiary questions followed. What expertise did they possess? What distinctions did they draw to include or exclude a particular issue? Which networks did the actor participate in or take for granted and which did they construct? How did their networking relate to the kinds of issues they wished to prioritise and the levels at which they chose to work? Thus to network is simultaneously to frame - the frame defines the limits of the network, and networking fills the frame. More broadly social actors produce and shape discourse, but also act within and are constrained by rules embodied in discourse (Giddens, 1984; Dryzek, 1997).

Dryzek (1997) has suggested a sequence of broad discourses through which environmental problems have been conceptualized. State regulators have felt most comfortable with the environment as a series of problems that may be solved piecemeal. Industry has been driven by a disregard for nature and environment, coupled with a Promethian belief in human ability to innovate in the face of biological problems. Environmentalism was launched in the 1970s with a Survivalist discourse, which posed environmental issues in Neo-Malthusian terms of overrunning natural limits. Later discourses have merged from the disputes between these existing discourses and encounters with social discourses of social justice, community and North South issues. Thus Green Radicalism is a less elitist descendent of Survivalism, while Sustainable Development tries to square the circle of maintaining capitalist development, social justice and environmental protection at the same time. In chapter two the Promethian ethic of patenting is set against a Green Radical critique. Chapter three traces the global expression of Promethian discourse in the WTO and the links between it and the global organizations dedicated to Sustainable Development, the FAO and the CBD.

Researching social movements raises political and ethical issues, as well as methodological ones. A possible political purpose of social movement research is that it can open up civil society, playing a part in the process of social reflexivity, alongside social movements themselves, requiring open and contractual research ethics (Melucci, 1992b; 1996). Methodological attention has to be given to the fact that NGOs and other social movement actors are centrally concerned with transforming social reality and so are intensely reflexive, often sharing a conceptual language with academics – using terms such as globalization, regimes and civil society. My intention in this book is to make a small contribution to this reflexivity.

2 Patenting Discourses: Enclosure and Expert Systems

> A patent right, because it gives its owner a monopoly, is the form of intellectual property par excellence (Bainbridge, 1992: 7).

> Patents are a paradise for parasites...patents protection forms a stumbling block for the development of trade and industry...the patent system is a playground for plundering patent agents and lawyers. (J. Geigy-Merian, founder of Ciba-Geigy Inc, 1889, quoted in Hobbelink, 1991: 99).

> It is Ciba-Geigy's position that legal protection of intellectual property serves the public interest by stimulating continuing investment in technological innovation (Ciba-Geigy executive, John H. Duesing, 1989, quoted in Hobbelink, 1991: 99).

Intellectual Property Rights (IPRs) attempt to parcel up creative expertise into discrete units that are economically and legally manageable. Patents deal specifically with technological expertise. The unique and interesting thing about patents is that they are both a knowledge claim and a property relation. This dual status means that the power lines of two expert systems - science and law - always traverse any patent. Patents act as stabilising points in these expert discourses, legitimating knowledge claims in terms of property relations and property in terms of creative knowledge.

The whole development of biotechnology as a new technology has gone hand in hand with an attempt to extend patenting into the novel territory of genetics. Unlike nuclear power, which was born of the post war social democratic consensus, biotechnology has bloomed in the harsher climate of the neo-liberal privatization and structural adjustment. In this period which has simultaneously seen the globalization and restructuring of the economies of the world towards knowledge-intensive forms and decisive shifts in the balance between the private and public sectors, intellectual property is becoming ever more important (Lash and Urry, 1994). The controversy over genetic patenting centres on whether it is ethical to extend a strong form of intellectual property from its mechanical industrial heartland to living organisms. Contentious areas include material taken from human bodies, natural organisms and processes, and seeds. Plant genetic resources (primarily seeds), generate the crops that feed the human population as well as provide medicines, clothing and other resources. Advocates claim that genetic patenting is an extension of the rational system for ownership of technology required for the development of the new dynamic biotechnology, with great potential for human welfare. Critics, however, view genetic patenting as an attempt by the biotechnology industry of the rich North to capture the gene wealth of the otherwise poor South without compensation. This chapter attempts to expand this critical claim by analysing the dominant discourse of patenting and situating it in terms of the expert systems that sustain patenting and allow the translation of an enclosure of property into an apparently rational system of bureaucratic regulation.

> A discourse is a shared way of apprehending the world. Embedded in language it enables those who subscribe to it to interpret bits of information and put them together in coherent stories or accounts. Each discourse rests on assumptions, judgements, and contentions that provide the basic terms for analysis, debates, agreements and disagreements (Dryzek, 1997: 8).

Central elements in a discourse include: basic entities constructed; assumptions about natural relationships; agents and their motives; key metaphors and story lines (Dryzek, 1997). This approach to discourse analysis defines discourses in a more or less linguistic way, similar to language games (Wittgenstein, 1967), rather than a wider concept of discourse that includes institutional settings and technologies (Foucault, 1977, 1978). This wider context is more clearly conceptualized as 'expert systems' (Lash, 1994) in that the discourse of expertise, and the experts themselves, are embedded in professional institutions and linked to technologies. Thus expert systems have been defined as 'systems of technical accomplishment or professional expertise that organize large areas of the material and social environments in which we live today' (Giddens, 1990: 27). Daily life requires placing considerable trust in these expert systems to provide reliable information and judgements and to regulate risks. Expert systems are bodies of knowledge practices, which are specialized, professionalized and stable. Legal recognition of scientific expertise requires that it be 'consistent, methodological, cumulative and predictive' (Kenny quoted in Campbell, 1989: 268 n 3). By applying apparently impersonal and testable criteria expert systems are able to disembed knowledge from any local social conditions and transfer it to the global.

The dominant discourse of genetic patenting hangs on a simple story line. Social wellbeing depends on technological progress. Such technological progress will only happen if innovation is financially rewarded. Therefore human wellbeing is best promoted by protecting corporate profits through a strong form of property rights over genes and seeds, namely patents. The appropriateness of patents depends on the use of mechanical metaphors to make organisms resemble machines. This dominant discourse is embedded in a combination of technoscientific and legal expert systems, including technologies, professions and institutions.

A critical discourse challenges patenting with its own story line and metaphors. The critical story line is that patenting shifts common property into private hands, causing poverty and restricting rights for the economic benefit of the biotechnology companies. Since patents control intellectual property, it is traditional, indigenous, local knowledge that is captured by experts and represented as their own. Compelling metaphors are generated, such as biopiracy and 'the enclosure of the commons'. The critical discourse is located in counter-expert NGO networks and their allies.

Patents as Knowledge Claims

The dominant discourse of patent law is a Promethian discourse. 'Promethians have unlimited confidence in the ability of humans and their technologies to overcome any problems presented to them – including what can now be styled as environmental problems' (Dryzek, 1997: 45). Promethian discourse subordinates passive nature to active human society, body to mind, tradition to expertise. As a legitimating discourse of rampant industrialization, it also subordinates farmers or citizens to high-tech corporations. Philosophically Promethians side with innovation and human creativity, but politically and economically patents side with the rich and powerful.

A grand narrative rooted in patent law is the absolute separation between the realm of the natural world and the realm of human endeavour. Patenting separates technological inventions as human constructions from the raw material of these inventions. Raw materials include the natural world and knowledge of this natural world in the form of scientific discoveries. Thus a central epistemological distinction is made in patent law between discovery and invention. Nature is discoverable through science, while the social realm is created by human beings in the form of technology or the arts. It is this second realm which comes about by human labour that therefore attracts property rights in the form of patents on technology and copyrights on artistic work. Discoveries as such are not protected, though the form in which they appear as journal articles carry copyrights. While these discoveries are claimed by individuals and require acknowledgement and are thus financially valuable in career terms to their authors, they do not carry any power over the subsequent use made of the content of the discovery. Indeed scientific papers are often conceptualized as contributions to public knowledge. In a copyright it is the way things are written not the usefulness of the information contained which is at issue. The copyrighted text is a narrative of who did what and it is the particular narrative that is owned. A patent on the other hand controls all possible embodiments of the knowledge contained.

Yet this modernist split between the human and the natural (Bauman, 1990) has been questioned by twentieth century philosophy. Scientific discoveries have been viewed as 'metaphoric re-descriptions of nature' (Rorty, 1989: 16) and hence creative human acts of language making, and not as is implied by patent law the uncovering of objective truth given by nature. Science is then recast in a creative, technological mould, thereby erasing the distinction between scientific discoveries and inventions. Latour (1987; 1988; 1993), in his studies on science and technology, has argued that the clear distinction between the two is untenable, substituting the term technoscience to describe a continuous field of activity. Thus the development of the scientific field of genetics, the technology of recombinant DNA and indeed the patent claims are moving together. Can the knowledge of the location of a particular gene (science) really be separated from the diagnostic test for it (technology)? Equally, when this research is done in the private sector (or even the profit-driven public sector) can this knowledge claim be separated from the property right it embodies? In effect the social conditions under which biotechnology operates has shaped research to such an extent that knowledge often only appears in a patentable form. Discoveries and inventions are best distinguished as products of competing genres of textual discourse. Scientific journal articles by definition reveal discoveries, whereas patents disclose inventions. The same authors may present the same ideas in both arenas, simply by rewriting to follow different discursive conventions (Myers, 1995).

The modernist conceptual distinction between culture and nature leads to a second major exclusion from patentability, covering natural organisms and processes; what are termed 'essentially biological processes'[1]. The basic argument here is that as naturally occurring processes or organisms these cannot be invented only discovered. This principle, however, has not prevented the evolution of IPRs over seeds. Plant Breeders' Rights grant a monopoly over particular plant varieties. Furthermore 'microbiological processes' are patentable[2]. The distinction between essentially biological processes and microbiological processes is obviously a fine one when discussing the transfer of genes between bacteria and plants and animals. The larger point though is

---

1 Clause 53B of the 1973 European Patent Convention 'excludes from patentability "plant and animal varieties" and "essentially biological processes", but not microbiological processes and the products thereof' (King, 1992: 2).

2 Greenpeace have challenged patents on seeds in the European Patent Office on grounds of attempting to patent varieties, essentially biological processes and contravening morality. They won on the variety argument, but predictably lost the other two.

that whatever the apparent logic of the unpatentability of nature, where a market in genetic material has developed as in the case of agricultural seed, a pragmatic solution has been found whereby plant breeders have been able to acquire IPRs over plant varieties. Plant varieties have been tacitly accepted as human inventions, though the special quality seeds have of self-regeneration has been taken into account by limiting the extent of their enclosure in IPRs.

The discourse of patent law invents three types of entities. (1) Nature constitutes the raw material on which human activity can take place. (2) Basic scientific methods act as tools to give access to nature and do not attract property rights. (3) Human actors apply these scientific tools to the natural raw material using human intelligence to create new inventions. Only this third entity is an agent and therefore holds the intellectual property rights to his/her invention. In the case of John Moore, whose cell line was extracted from his spleen and patented, his body became raw material, whereas the doctors assumed the active agency of creative intelligence. If we apply the logic of the same argument to the case of a biotechnologist transferring a gene into a seed to produce a new plant variety several entities appear as passive raw material, and only one as an active agent. Both the recipient seed and the source appear as natural raw material, the new seed becomes a patentable technology only the biotechnologist who transferred the gene is an agent. The seeds that form the source for this innovation may well be an unregistered and undefined Third World variety, known as a 'landrace'. Farmers disappear altogether from the equation, even though they may have selected both the recipient seed and the donor seed for the presence of the relevant gene. Farmers (and gardeners) fail even to be registered as part of the public domain where relevant knowledge that could prejudice the grant of a patent already exists. Furthermore, agency in the patenting world shifts to an ever Increasing degree, from individual inventors to large companies able to finance research programmes and patent applications.

Mechanising Life: Cybernetic Metaphors

A powerful story line is required to convincingly justify patenting to previously natural and therefore unpatentable processes of reproduction of species. The argument that genetic patents follow logically from mechanical patents has been facilitated by the mechanical metaphors of genetics that have been deployed in the expert system of molecular biology. Mechanistic metaphors have became literalised in the rise to scientific power of molecular biology and are the basis of the claims of biotechnology to be enhancing the world or even generating a new world.

Molecular biology was launched by two major atomic physicists - Bohr and Schrodinger - who pushed for the establishment of a molecular basis for biology; that is, systematic physio-chemical explanations for the nature of living organisms. The subsequent history constituted a major victory for reductionism (Wheale and McNally, 1988). The Rockerfeller Foundation supported molecular biology from the 1930's as part of their integrated strategy for reductionist science encompassing physical, biological and human sciences (Busch, Lacy, Burkhardt and Lacy, 1991).

The experience of participating in the war effort in the 1940's had an immense impact on biologists. The wartime information systems research, with its emphasis on communication, command and control, produces cybernetics, which becomes the dominant paradigm in 1950's biology (Haraway, 1992). Within the metaphor of cybernetics, organisms are replaced with communication technologies. The direct impact of cybernetic thinking on molecular biology is outlined by Yoxen, who suggests that we are encouraged to think of:

> organisms as 'self-assembling, self-maintaining, self-reproducing information-processing machines'. The idea that the genes embody *encoded information*, which is *transcribed* into a chemical intermediate, *messenger* RNA, and then read *codon* by *codon* and *translated* into a *sequence* of amino acids to form a protein molecule is basic to the conceptual framework of modern biology. Genes are *switched on* and *off;* messenger RNA is *processed* and *edited;* gene sequences contain *signals* to the cellular *machinery*; and information is *read out* from the DNA (Yoxen, 1983: 44-5).

Molecular biology operates through information metaphors drawn from computing, cryptography and cybernetics. The title of a popular TV series on genetics in 1993 was *Cracking the Code*, an obvious reference to WWII signal corps deciphering enemy intelligence. This familiar image of heroism and technical skill gives molecular biology a powerful access point to popular culture.

Three central metaphors are used to legitimate agricultural biotechnology and patenting seeds. First is the reduction of biological processes to chemical ones. Second is the implicit model of computer language, sometimes simplified to 'a more publicly accessible analogy - a Lego Kit'. Third, 'naturally found organisms and their cells are explicitly reconceptualized as "factories" which have already evolved a productive efficiency, a quality to be enhanced by genetic engineering' (Levidow, 1991: 553). These three metaphors - the computer, the factory and chemistry - are mobilised by the agro-chemical companies to present themselves as 'Green and clean' by enhancing nature through their interventions. More recent advertising material portrays the use of biopesticides as a 'clean surgical strike' in the post Gulf War military metaphor (Levidow, 1996). The use of these mechanical metaphors has allowed biotechnologists to present their work in ways that are comparable with mechanical engineering with its established system of industrial patents. In the process it has granted the gene an autonomous existence free of environmental influence (not unlike the bourgeois individual). A number of theories exist on the role of genes, but at the least it is clear that the same gene will not express the same amino acids in different contexts or different organisms. Various types of genes have been described - structural genes, jumping genes -and a number of competing definitions advanced, yet there is no one unified definition of a gene that covers all cases (Kolleck, 1994). Leading to the claim that the meaning of the term 'gene' is far too complex and ambiguous to be the basis of the grant of a patent (Kolleck, 1995).

Knowledge in the Public Domain

A patent is a way of making an invention public, while simultaneously claiming the ownership thereof and thereby maintaining a monopoly of its use. What is already in the public domain is unpatentable. Hence a patent always involves a disclosure which puts some new information into the public domain. The disclosure of necessity precedes the grant of a patent, so that if a patent is refused the invention has still been disclosed and cannot therefore be removed from the public domain. For this reason biotechnology companies do not always patent their work, sometimes preferring the route of an undisclosed trade secret (Wheale and McNally, 1993). A more subtle method is to disclose only part of your work - enough to gain the grant of a patent without disclosing the direction of future research to competitors.

There are of course other ways of making an idea public without claiming a monopoly of its use. A published scientific paper is copyrightable, but would not prevent others from using the material disclosed. Underlying these two quite different processes of disclosure - publication or patenting - is the historic institutional divide between the university and the private sector. The division of labour between basic scientific research in the universities producing public knowledge and applied or near-to-market research in the private sector has broken down, with prominent university based biotechnologists setting up private companies with venture capital to exploit patents derived from their university research (Kloppenburg, 1988a). The distinction is blurring even more with the privatization of research institutes and the introduction of patenting targets into government research institutes (Interview 26/3/1994).

An academic article is intended to address a specialist in a particular area and the author claims authority by locating the article in a specialist literature, and hopes to make an incremental contribution to this literature. A patent by contrast, addresses a 'person skilled in the art', the author seeks to separate the patent from preceding texts to establish its novelty and attempts to claim wide ranging possible applications in any area of technology (Myers, 1995). While the article is central to the culture of science, a patent acts as a boundary object defining the continuities and discontinuities between science and law.

A patent can be granted on a product or a process and consists of two parts. (a) The "specification" which includes the background, a summary and an explanation of how to produce at least one embodiment of the patent. The explanation should be sufficiently clear to allow a reasonably skilled craftsperson to reproduce it. (b) The "claims" which includes only what is new in the patent. Both of these are judged in relation to the "state of the art" in the particular industry (Campbell, 1989). A hypothetical skilled worker should be able to make the patent as specified work and the new knowledge claimed should be beyond what this hypothetical skilled worker would be able to achieve through application of skill. It is often recognized in the grant of patents that the notional skilled worker is in reality a research team bristling with PhDs (Campbell, 1989). The public domain recognized by patent law is interior to the expert system itself, recognising only what is known in the industry itself or has been published in scientific journals. In the case of seeds this excludes all vernacular knowledge of the uses of locally cultivated landraces.

The criteria for the successful grant of a patent are widely accepted in legal systems internationally, although the interpretation of these criteria may differ markedly. A patent application must fulfil three criteria - novelty, inventive step and industrial application. These criteria spell out in more detail the relation of the patent to the accepted state of the art.

The question of novelty is addressed in the UK Patent Act of 1977 which states that 'the invention is new if "it does not form part of the state of the art"' (Bainbridge, 1992: 264). The state of the art includes anything which is in the public domain 'whether by written or oral description, by use or in any other way' (Bainbridge, 1992: 264). Nothing, which is already in the public domain, can be patented.

Traditional cures used by Kayapo Indians in the Amazon might be public knowledge among the Kayapo, or known only by an expert monopoly of local shamans. Either way they would not count as being in the public domain for the purposes of the US Patent Office. Effectively the public domain referred to in patenting discourse is a theoretical space interior to the expert system concerned. Hence the novelty of the process of isolating a chemical compound or the expression of a gene is judged in relation to other similar processes or products as defined within the same discourse. The vernacular knowledge claims of the Kayapo are literally, but more importantly figuratively, in a foreign language, which is not recognized by the patenting authorities as knowledge at all. This gives rise to a 'differend' or dispute in which the opposing sides do not agree common criteria for a judgement one way or the other (Lyotard, 1988). The reductive pharmacological model of isolating active ingredients of course also operates against holistic medicine practitioners in the West, whose claim to have developed cures are only taken seriously to the extent that they can be translated into the language of active ingredients. Even then, ownership would presumably go to the scientists isolating the compound, not the practitioner using the whole plant. In Switzerland a patent was applied for on all future varieties of camomile with higher than average levels of active ingredients. The Swiss Association of Organic Herb Growers successfully opposed the application, partly on the grounds of lacking novelty (Baumann, 1995: 14-16). However, in general it would seem, local varieties of cultivars developed by farmers do not constitute part of the public domain, and only enter the world of intellectual property rights when collected by western scientists and thereby disembedded from the local conditions and transferred into a global expert system. Thus it may be argued that a patent does not necessarily involve new knowledge, but can simply refer to the enclosure and transfer of knowledge from one cultural context to another. The earliest meaning of novelty in British patent law referred simply to the introduction of a technique to the realm without regard to whether it had been copied from foreign users (Campbell, 1989: 212). The cultural boundedness of patenting remains, disguised by the subsequent development of more universalising knowledge claims.

An inventive step is thought to have occurred when 'the invention is not obvious to a person skilled in the art' (Bainbridge, 1992: 267). Again this innocuous craft definition of creativity covers an important distinction in knowledge-power. Even when collected, local farmers' seeds are not viewed as plant varieties, but as landraces or 'primitive cultivars' and grouped with wild plants as 'exotic germplasm' - genetic material falling outside the expert system. Farmers' varieties are thus classified as raw material for scientific plant breeders, conflating farmers' skilled labour with natural selection. The separation of individual creativity from tradition is difficult enough where there are agreed cultural parameters. At a global level it generates a politically and ideologically charged relationship of unacknowledged intellectual debt between first world experts and third world farmers, transferring credit for the skilled work of the farmers to western individual experts and corporations.

A patent must have an application in an industry including agriculture[3]. Again the classification of agriculture as an industry raises questions about subsistence agriculture. A patent application is a written document, and to be

---

3 A therapy or diagnosis is not counted as a suitable application for patenting, but drugs used in therapies may be patented (Bainbridge, 1992: 270).

successful must be deemed to be sufficiently clear and instructive to be carried out by the 'skilled worker'.

The problem of defining which knowledge falls in the public domain and which does not has been graphically illustrated by the case of the Neem tree. Various parts have been used in India for medicinal purposes, not least as a common way of brushing one's teeth. Transnational corporations have taken out patents on derivatives of the Neem. Precisely by a reductive approach they are able to claim that the derivatives they have produced are novel patentable inventions, excluding the everyday practice of millions of Indians from the public domain and enclosing their communal Neem tree for corporate profit (Pearce, 1993; Shiva, and Holla-Bhar, 1993).

Patents are an industrial form of property, which previously excluded living matter from their reach. They can only be invoked through high-tech processes involving expert discourse on genetics. Hence if local people use a whole plant which they have developed through selective breeding it is not patentable, but extracting a gene and using it in another organism is. Patenting cannot be separated from the industrial and scientific knowledge form. It applies to certain types of knowledge claims and not to others. It is not about making a plant perform a particular function, such as the Neem twigs cleaning your teeth, but about an engineering knowledge claim, concerning for example the extraction of a chemical compound or isolation of a gene. The patent application must be novel and inventive in the eyes of chemists and biotechnologists, and clearly and instructively written for them not the current users of the Neem tree. Patent law draws a boundary between the experts whom it addresses and the lay population, whose knowledge is excluded as invalid.

Ethics: Patent Justification Theories

Beneath the case by case justification of patent claims lie philosophical discourses that are mobilized as a justification of why knowledge should be private property. The argument for patents has shifted over time. Thus the oldest argument is that property is a natural right.

> A man should own what he produces, that is, what he brings into being. If what he produces can be taken from him he is no better than a slave. Intellectual Property is, therefore, the most basic form of property because a man uses nothing to produce it other than his mind (Bainbridge, 1992: 17).

Historically, there have been variable views on this argument. The 1791 French Patent Law called the patents the natural right of the inventor, while the 1794 Austrian Patent Law, claimed that patents form 'an exception to the natural right of citizens to have access to inventions' (Raghavan, 1990: 115-6). Bainbridge summarizes four theoretical justifications for patents put forward in the nineteenth century - contract, reward and incentive theories as well as natural rights. The natural rights theory simply asserts the individual's property right over his/her ideas that should therefore not be stolen. The contract position advocates a deal whereby temporary protection is granted in exchange for new knowledge, the reward system argues that inventors should be rewarded for making useful inventions. The incentive argument carries this further. 'By constructing a framework whereby invention is rewarded, this will act as an incentive to make new inventions and invest the necessary time and capital' (Bainbridge, 1992: 248).

In surveying contemporary justifications for patents, Ko (1992) came up with four variants on the incentive schema - Incentive to Invent, Incentive to Disclose, Incentive to Innovate and Prospect Theory. In each the monopoly profits or rents accruing to the patent holder is justified in different ways as promoting the public good. In Incentive to Invent theory 'an inventor demands compensation for his investment in research and development' (Ko, 1992: 791) and therefore the monopoly 'price reflects value to the invention's users rather than the mere cost of production' (Ko, 1992: 792). According to Incentive to Disclose theory 'without patent protection inventors would conceal their inventions in order to prevent exploitation by competitors.' Incentive to Innovate theory 'recognizes that inventions may require considerable further investment beyond mere discovery for commercial exploitation.' While Prospect Theory maintains that 'patents promote efficient development of patented inventions by allowing patent owners to co-ordinate further research and development.' (This last theory owes its name to the prospectors of the Wild West with all the disregard for common property that implies. This does not augur well for debates over the sovereignty over seeds.) Some authors are so firmly convinced of these theories as to feel able to publish tables detailing the numbers of products which would not be developed if patents did not exist (Mansfield 1986 reproduced in Nogues, 1990: 87). Pharmaceuticals heads the table followed by chemicals. These theories taken together (for their differences are less important than their similarities) with their claims that ownership leads to invention, disclosure and innovation of technology assume that these in turn lead to economic growth and increased human welfare. The whole argument that patents lead to economic growth was found to be unprovable by a US Senate study (Matchlup, 1959, quoted in Yamin, 1993: 58 n 120). The broader point that economic growth is itself an undeniable advance in human welfare is of course now heavily contested by the Green movement.

Patents are enshrined in the American Constitution on the grounds that they 'Promote Progress of Science and useful Arts' (Art 1, 8, cl 8 of the *Constitution of the United States of America*, quoted in Ko, 1992: 79). This notion of progress forms the bedrock of all the discussions about patenting. While disputes may occur over the scope, nature or applicability of patents in different areas those who advocate patenting never question the paramount importance of scientific and industrial progress. Hence Ko concludes his discussion of patent scope with 'The proper scope is that which best promotes technological progress' (Ko, 1992: 804). Yet progress, as an intellectual foundation, or grand narrative of modernity, has been under attack both philosophically (Lyotard, 1986; Bauman, 1993; Touraine, 1995) and politically from the environmental movements.

Two of the earlier arguments in favour of patents may be derived from what Touraine (1995) calls the two faces of modernity: 'democratic' and 'republican'. The claim of a natural right of an individual to the products of his/her mind draws on the democratic theme of the enchantment of the subject. Here the work of John Locke and his influence in early American thought is particularly important. On the other hand, appeals to progress depend on the historicist identification of the individual good with social utility. Progress equals the good of the republic to which the individual must submit. Indeed the US Supreme Court decision on the John Moore case may be seen as a decision in favour of progress as defined by the biotechnology industry over the democratic rights of an individual.

The philosophical arguments about natural rights and progress now form a background to more pressing strategic economic concerns over reward for investment. The shift in emphasis to the investment argument reflects the fact that big companies own most of the patents in the world. It is these big industrial players dominating the global markets that utilise patents to retain monopoly rents from the technology they own for as long as possible. What is significant about the multiple arguments in favour of patents is the layering or sliding of one position to another. So that the ethical force of the natural right of individuals to own their own thoughts, and social progress through rationalization, are harnessed to quite different legal personalities, that is transnational companies wishing to protect their investments. This sliding of individual rights to the protection of corporate wealth is a central ideology within American capitalism as a whole. Such an ideology is bound to exist to excess (Sartre in Versfeld, 1991) that is, beyond that which is necessarily of functional value to the key players. This always allows criticism of the stupidity /irrationality/inefficiency/mindless bureaucracy of a system, which produces side effects that are harmful without any benefit accruing to the major players the policy is intended to protect. Hence the loss of many of the smaller varieties of vegetables in Europe, due to the registration system, is of little direct benefit to the seed companies. This ideological excess also allows for the ideology to be used in a variety of ways. Patenting can, for example, be used to motivate employees, rather than to produce tangible benefits to a company (Campbell, 1989). It should be noted that a primary purpose of patents is to maintain an advantage against other similar actors. Hence several transnational agrochemical companies are currently locked in litigation with each other over alleged infringement of each other's patent on genes. The ability to produce the insecticidal toxin Bt has been genetically engineered into various crops by competing companies, in order to produce varieties resistant to insect pests, resulting in competing patent claims. Among others DeKalb has sued Pioneer, Mycogen, Ciba and Northrup King (a Sandoz subsidiary), while Monsanto has sued both Mycogen and Ciba. Similarly, PGS has also sued Ciba and Mycogen and DeKalb has sued two small regional seed companies for selling Northrup King's Bt corn (GeneEthics News, 13, Sept/Oct 1996).

Economic Reach: Patent Scope

An interface between the patent as legal property and as technological expertise is expressed in the scope of the patent. In other words the range of products or processes that would fall under the monopoly held in a particular patent. 'The economic power a patent confers depends on its scope' (Ko, 1992: 778). In determining this scope the US Patent Office employs the Doctrine of equivalents. Pioneering innovations are thought to deserve wider scope of patent than more mundane inventions. If the scope is too narrow endless almost identical patents can be filed. If on the other hand they are too wide a single patentee can hold the rights to whole areas of technology. For example, Boyer and Cohen in 1973 acquired the US patent for rDNA technology for Standford University. As technology has become more complex - patent applications have tended to become heavily coded so as to disclose the minimum information and to lay claim to the widest possible technological territory. An application generated by the Human Genome Project involved 2000 overlapping claims (Jones, 1992: 457 n 9).

Patents as Property Claims

Intellectual property, like other property, can be bought and sold. The commodity defined by a patent is the monopoly of use of knowledge over a period of time. Intellectual Property can also generate rent - the patentee may licence third parties to produce a patented product in exchange for royalty payments. In the case of seeds a farmer may be expected to pay annual royalties on the seeds for re-using them to grow crops. Patents are both public knowledge and private property. While advocates of patenting make much of their importance as public knowledge, it is their role as private property that is dominant. Scientists regulate scientific papers as public knowledge through peer review. On the other hand, law courts, within the wider context of governmental policy decide the validity and scope of patents. While scientific experts may give evidence, the decisions rest ultimately with judges who deploy a kind of lay pragmatism, often informed by crude economic arguments about the public good and progress.

Applicants for patents have recourse to a sequence of experts. First in line are patent agents - scientists trained in patent law in practice, who file patent claims on behalf of their clients. Patent agents may represent their clients in the lower levels of the judicial system such as patent offices and county courts. Barristers handle disputes in higher courts and may appear in hearings at the European Patent Office or the Patent Court. Judging patent cases has become an increasingly specialized business. Juries for civil cases were abolished at the turn of the century, then with the establishment of the English Patent Court in 1949 circuit judges were replaced by specialized patent judges. This has in turn changed the nature of expert evidence required by the courts, as the judges are more familiar with the issues involved in patent cases. While patent agents may still appear as expert witnesses, barristers have found it more effective to call witnesses who are practitioners to discuss the clarity of a patent claim and help define the current 'state of the art' and hence support a claim of novelty. Less frequently research experts are used to demonstrate distinctiveness of techniques (Campbell, 1989).

Patent agents attempt to get the widest scope possible included in their client's claim, as it is the scope that determines how many future innovations will fall under the patent. In the early period of biotechnology they were extremely successful, even though few patent agents were trained in biotechnology, with chemists usually handling biotechnology. In the early 1980's biotechnology provided a new growth area to patent agents that had become becalmed as a profession. The biotechnology boom subsided by the mid-1990s, as claims awarded have become more modest. This follows a general pattern in which earlier innovations in a particular technical area are granted wider scope than subsequent claims. Whatever the rationality of this process, in the end it comes down to whether the patent agents and barristers can convince the patent offices and judges of the validity of their claims (Interview with patent agent, 5/6/95).

Patents by Precedent

A series of US Supreme Court decisions have set precedents that have steadily eroded the exclusion of genes from patenting. In 1980, the Supreme Court ruling in the landmark case of Diamond v. Chakrabarty found that a genetically engineered bacterium could be patented. In 1985 the US Patent and Trademark Office granted a patent for a variety of maize. Then in 1988, the first patent was granted on a genetically engineered animal - the

Oncomouse, a laboratory mouse, owned by Du Pont and Harvard University, which has had a genetic tendency to breast cancer introduced into it by rDNA. In 1994 the patent on a human cell-line was upheld by the Supreme Court. In a bizarre sequence of events, American construction worker, John Moore, had his diseased spleen surgically removed. He subsequently discovered that the National Institute of Health was patenting the cell line from his spleen. He challenged the patent, but the Supreme Court Judge argued that to allow his claim to ownership of all products derived from his body parts would present obstacles to progress.

> [If the] plaintiff is permitted to have decision-making authority and a financial interest in the cell-line, he would then have unlimited power to inhibit medical research that could potentially benefit humanity (*Ecologist CAMPAIGNS*, p 2, *Ecologist*, 24 (5) Sept/Oct 1994).

This begs the question of whether ownership of the same cell-line by a biotechnology company is not an equal obstacle to potentially useful research. Judges make key legal decisions over patentability. Their judgements are made as to what is in the public interest, based mainly on the arguments put forward by the biotechnology industry, which confuses its own profitability with the interests of wider society.

The situation has developed more slowly in Europe. There has been the same gradual erosion of the notion that living organisms cannot be patented, but not without some reversals for the pro-patenting lobby. The European Patent Office (EPO) in Munich administers the European Patent Convention. It tends to follow American precedents, but it allows objections on moral or ecological grounds. Both moral (e.g. animal welfare) and ecological objections are supposed to be assessed by balancing risks and benefits (Noiville, 1994). In 1989, EPO granted a patent for a 'technique for enhancing seed storage protein' (King, 1992: 3). Transgenic animals joined the list of patentable subject matter when the EPO granted the Oncomouse patent on 13 May 1992 (Thurston, 1993). This decision was very controversial, overturning a previous decision to refuse the patent in July 1989. A number of oppositions were registered and the Greens in the European Parliament succeeded in passing an emergency motion in February 1993 calling for the repeal of the Oncomouse patent. The decisive 178-19 majority (Thurston, 1993) indicates a strong European resistance to simply being drawn along in the wake of the US, although the EU and therefore the European Parliament has no jurisdiction over the European Patent Convention, which is a separate international regime including non-EU members such as Switzerland. The result of the Oncomouse decision was 'about 5000 patent applications for genetically engineered plants and animals being processed by the various national patent offices of Europe' (Wheale and

McNally, 1993: 266). Within the framework of these precedents countless genetic patents have now been filed, with well over 2000 field tests carried out in the UK alone. Notorious patent claims on plants include all types of genetically modified cotton, soya and non-squishy tomatoes.

Patents as Public Policy: The European Patenting Directive

Patent law has historically been part of domestic property law and also part of national technology policy. There has therefore been considerable variation over time and between countries, in the period the patent runs for, the limits of patentable subject matter, and the conditions under which governments may intervene to compel patentees to issue licences in different countries. In addition, 'All patent systems designate certain no-go areas as a consequence of public policy considerations' (Cook, Doyle and Jabbari, 1991: 115). There has, once again, been considerable variance in what these no-go areas are, since they are politically defined. Many countries, such as India, have excluded food, agriculture and pharmaceuticals, since they are seen as basic to public welfare. The project of homogenising patent law globally is the subject of the next chapter, but in Europe closer regional integration already existed through the EPO. The European Patent Convention clause 53A excludes inventions that are contrary to *'ordre public'* or 'morality' and clause 53B excludes 'essentially biological processes' as well as plant varieties.

Frustration with legal regulation through the EPO led European biotechnology companies to want to change the legal framework. Inexplicably, rather than attempting an overhaul of the European Patenting Convention, the biotechnology industry chose instead to launch a major offensive in the late 1980's via the European Union, in spite of its lack of jurisdiction over the European Patenting Office. Thus they opened the issue to debate and opposition in the European Parliament. Acrimonious debates surrounded the European Draft Directive on Patenting Biotechnology throughout its seven-year life. Published in October 1988 by the European Commission, following intensive lobbying by the biotechnology industry, the Draft Directive (Proposal for the Legal Protection of Biotechnological Inventions) was to patent all life forms. Hence article 2 states that 'a subject matter of an invention shall not be considered unpatentable for the reason that it is composed of living matter' (Wheale and McNally, 1993: 530). Articles 3-8 expand on this principle. Secondly, it holds that to reproduce patented living organisms without the consent of the patent holder infringes the patent (articles 9-11). The fact that living organisms reproduce causes a clash between two rights normally recognized in patent law. The right of a patentee to control the multiplication of what is claimed in the patent clashes with the right of a consumer to use an invention for the purpose for which it was intended once purchased. The directive gives precedence to patentee over the purchaser (Roberts, 1995). That is, the seed companies would retain rights over subsequent generations of seeds. Therefore, for farmers to grow out patented seeds would require them to pay royalties to the seed company every year.

The draft directive ran into considerable opposition in the European Parliament and so it bounced back and forth between plenary sessions of the Parliament and various committees, particularly the Legal Affairs Committee, and the Commission, generating more than 50 amendments (Jones, 1992: 455). The Earth Summit in June 1992 brought fresh troubles for the struggling draft directive as opponents suggested the directive would conflict with the Biodiversity Convention signed in Rio. The draft directive cleared this hurdle but amendments on farmers' privilege and PBRs were incorporated (Jones, 1992: 455 n 3).

On 16 December 1992 the European Commission published the amended draft directive (Thurston, 1993: 187), but the new draft did not clear up several problems. Firstly, it attempted to refuse patents for the human body or body parts per se but allows human genes to be patented if linked to a process or function. However, genes can obviously be viewed precisely as human body parts. Secondly, processes that transform the genetic identity of humans for non-therapeutic reasons could not be patented. Nor could processes that are contrary to the 'dignity of man.' None of these terms were precisely defined and in Thurston's view caused as many problems as they solved. Thirdly, Farmers' Rights were introduced into the draft directive. By this it was meant that farmers may breed from patented plants and animals on their own farms, but would not be allowed to sell seeds or animals for breeding purposes. Furthermore, co-ops would be excluded from sharing these patented organisms without breaching the patent. Meanwhile, the Commission remained resolute in its rejection of farmers' privilege as such, by which they meant the complete control by farmers of the plants and animals produced on their farms. In short, the Commission produced a series of fudges and compromizes in order to placatc the Green and farmers' lobbies in the European Parliament, while attempting to carry through this pro-biotechnology piece of legislation. The draft directive also faced the legal question of whether the EU has the right to modify the European Patent Convention, which is clearly a separate organization with non EU members (Thurston, 1993). Extra pressure was applied through the Senior Advisory Group on Biotechnology (SAGB), an industry lobby group, including ICI, Monsanto, Hoechst, Sandoz and Unilever. SAGB claimed that European biotechnology was in an unfavourable position compared to US and Japan encouraged the EU to produce a Communication in 1993 promoting biotechnology, known as the Bangeman Report. It specified biotechnology as a key technology for European economic growth. One of its proposals was to strengthen intellectual property rights by rushing through the draft patenting directive.

By the end of 1994, the Commission and the Parliament were deadlocked on six points, including farmers' rights. However, the most crucial point was the precise wording of the clause on the patenting of human body parts. The Commission favoured excluding only human genes in the 'natural state in the body' (Roberts, 1995b: D-117)[4] while the Parliament was adamant that they opposed the patenting of human genes 'as such'. The distinction being that in the Commission's position it was possible to patent genes from a blood sample once it has been taken, whereas in the Parliament's view such patenting was precisely what they were trying to avoid. This led the patenting directive to gain the distinction of being the first directive to undergo the conciliation process laid out in the Maastricht Treaty. Twelve MEPs led by a German Social Democrat were expected to negotiate with the Commission. The rules for the conciliation process set out two different voting requirements for rejection of the results of the conciliation by the European Parliament. If no compromise was reached the directive could only be defeated by a two-thirds majority of the Parliament. On the other hand, if a compromize was reached, it could be rejected by a simple majority of the Parliament. Either way a rejection would lead to the directive being thrown out altogether. In the estimation of a group of NGOs[5] their best hope lay in a bad compromise being agreed. The advantage of a bad compromise was that it would be unacceptable to the Parliament, and because it was agreed, it would only require a simple majority of MEPs to reject it. On their knowledge of the MEPs involved in the conciliation procedure, the NGOs decided to leave them to make as bad a compromise as possible with no lobbying. At the beginning of February 1995 just such a bad compromise was announced triggering a rapid lobbying by a range of NGOs. The key problem was to separate the socialist bloc of MEPs from their lead negotiator on the directive. This was achieved and on 1st March 1995 the directive fell. The human gene issue was probably crucial as it allowed the mobilization of the churches and medical lobbies (Interview, 27/6/95). Greenpeace called it 'a moral victory for nature'. However, barely a year passed before a new patenting directive was on the move again this time being successfully passed in July 1997.

---

4 Unofficial translation from the French by the author.

5 This sequence of events and the thinking of the NGOs at a meeting in Bern on 22/10/94 were reported by Genetics Forum (*Splice of Life,* 1995). I also attended the meeting myself and noted in particular how only one person, Linda Bullard, the Genetic Policy Adviser to the Green Group, really understood the intricacies of the conciliation process, yet their adversaries seemed to lack even one person able to grasp the effect of the process on likely outcomes.

Plant Breeders' Rights and F1 Hybrids

There are a number of different forms of IPRs, which apply to seeds – patents, Plant Breeders Rights and trade secrets. Genetically Modified Organisms (GMOs) and gene splicing techniques have been the subject of patent claims. Plant varieties may be registered with Plant Breeders Rights (PBRs) or rights to maintain a variety on the EU register of legally tradable varieties. Hybrid seeds may in addition be sold while their parental lines are kept as trade secrets. The effect of the patenting debate has been to legitimate the strengthening of other forms of intellectual property, most notably PBRs so that there is less and less to choose between patents and PBRs.

Plant Breeders' Rights (PBRs) can only be attached to a plant variety, whereas a patent may cover both larger and smaller taxonomic categories of both plants and animals, as well as genes, micro-organisms and 'non-biological processes'. Historically, PBRs have granted more limited rights to breeders than patents, but the effect of the patenting debate has been the tightening up of PBRs to practically the same level of stringency as patents.

PBRs were first introduced in the USA, with the Plant Patent Act of 1930, and later internationalized in the Convention of the Union for the Protection of New Plant Varieties (UPOV) in 1961, updated in 1978 and 1991. Following the formation of UPOV, the UK passed the 1964 Plant Varieties Act, which set up a national seed register listing each variety together with the breeder who owns the PBRs to the variety. The 1964 Act was part of a longer governmental concern with seed quality, a concern arising from farmers for a guarantee of the quality of the seed they were sold; including that the seed sold was of the variety advertized. This required that the varieties be registered to prevent unscrupulous seed merchants from simply changing the name of a variety as a sham claim of novelty.

In order to get a new variety accepted onto the national list a breeder must pay the Government £2000, spread over 3 years, for tests to prove the seed does in fact produce plants that meet the criteria of a plant variety, namely distinctiveness, uniformity and stability. To maintain a variety on the list, a breeder must pay an annual fee of £350 (Cherfas et al, 1996). Varieties with a small volume of sales are unable to generate the profits to justify paying the registration fee, leading, initially to their exclusion from the market, and all to often, to their subsequent extinction. The formation of an EU Common Catalogue of plant varieties has further reduced the varieties available in Europe by deleting those deemed to overlap or not meet the standards of an UPOV plant variety, subject to PBRs (Vellve, 1992).

For a type of plant to be recognized as a variety able to attract a PBR it must be distinct, uniform and stable. Landraces and heritage varieties frequently fail these criteria because they contain too much genetic diversity - a distinct advantage in peasant agriculture attempting to confront varied conditions, but a disadvantage in acquiring exclusive property rights. It is this genetic diversity which allows individual plants within a particular sowing to grow at varying rates, fruit over an extended picking season, and survive adverse conditions to a variable degree. All of which makes them ideal for gardeners and small subsistence farmers, but problematic for large-scale commercial growers who need to pick a whole field on one day, by mechanical means wherever possible. PBRs are geared to large scale commercial agriculture and tend to exclude any alternatives.

The 1991 update of UPOV granted new rights to plant breeders at the expense of farmers. The 1978 version of UPOV contained the concept of farmer's privilege, which recognized the rights of farmers to save and re-use seed grown on their own farms. This farmer's privilege has been downgraded in the 1991 version; allowing seed companies to collect royalties annually from farmers re-using seed. UPOV 1991 also includes the extension of PBRs to 'essentially derived varieties' allowing breeders to claim royalties on varieties that share key genetic characteristics with the variety they own, even if those characteristics were derived from a local landrace (McDougall, 1995). Hence the work done by farmers to adapt a commercial variety to local conditions does not avoid royalty payments on the parental line. Conversely, a company that develops a landrace into a commercial variety could force farmers using the original landrace to pay royalties on the grounds of the genetic similarity between the two varieties. Unlike the original landrace, the new variety may be registered for a PBR. In other words the same issues arise as with patented seeds, even without the genetic engineering required for a patent.

The conditions agreed by the 13 member states of UPOV in 1991 are likely to be enforced on the much larger membership of the WTO via the TRIPs as the only alternative to patents (see next chapter). The distance between patents and PBRs is now very small. Their evolution displays the integration of technoscientific knowledge with property relations that is a central feature of current biotechnology.

A very effective way of guaranteeing IPRs in seeds is to inscribe them in the biology of the seed itself. F1 hybrids are the result of the first crossing of two varieties. They display 'hybrid vigour' but if seeds are collected from the crops grown from hybrid seed, they will not 'breed true'. That is, only a percentage of these second generation F2 hybrids will display the same characteristics, while in others less desirable recessive traits will emerge. In some commercially very important cases, such as maize, the F1 hybrids are actually male sterile, so they are unable to produce viable seed at all. The decision to base the breeding programme of a species on hybrids is not dictated by botany (Interview 28/3/94). Rather, it is a way of internalising the royalty payment into the price of the seed and forcing the farmer to pay for new seed every year. The knowledge embodied in the seed remains with the company that owns the 'elite' parental lines used in the cross that produces an F1 hybrid. The genetic makeup of the parental lines may then simply remain undisclosed as a trade secret.

Hybrid seeds provide the perfect formula of self enforcing IPRs. Breeding on from them is economically very risky in the short term at least, if not impossible. Patents and PBRs, if used on open-pollinating varieties, require enforcement. Enforcing the payment of royalties from farmers who re-sow their seed requires the collection of data about what farmers are actually doing. As ever, a variety of surveillance techniques are available: written records, high tech satellites and of course, informers. Satellite surveillance is particularly prevalent in the USA where fields are big (meeting, 21/6/95). In the EU subsidy application forms are thought to be the most efficient way of monitoring what farmers grow. Seed dressing companies can also be required to produce records of what seeds they have cleaned on which farms.

Critical Discourse: Enclosure of the Commons

A significant literature exists dealing with genetic patenting as a commodification (Kloppenburg, 1988a) or an enclosure of the commons (Shiva, 1993). These metaphors invoke the arguments of Marx on the primitive accumulation of capital (Marx, 1976). In this early phase of capitalism, capital is not accumulated primarily in the form of the extraction of surplus value from the proletariat but is simply taken from the peasantry by force. Under feudal relations of production the land is the property of the feudal lord, that is s/he can extract rent from the peasant tenants, but the peasants have possession or the right to use the land (Balibar, 1970: 212). In addition to their individual tenancies they have rights of access to common land. During the enclosures the peasantry are separated from land (i.e. their means of production) and consequently proletarianized. Marx recounts how in Britain the feudal lords transformed their property rights over the land to include exclusive rights of possession, both of the traditional tenancies of the peasants and the common land.

Bauman suggests that appropriation of pre-capitalist or pre-modern resources, including ecological and social arrangements, is a permanent requirement of modernity, and central to the self- destructiveness and injustice of modernity (Bauman, 1993). From the enclosure perspective, ecological problems, such as rain forest destruction, are largely traced back to this violent appropriation of the land (Hecht and Cockburn, 1989) and the removal of local rights of usufruct to local resources. The loss of informal structures of rights to use commonly held resources in the Third World are the cause of immiseration of the local poor, in the same way as Marx documented in sixteenth to nineteenth century Britain (Marx, 1976). Larger ecological claims can be made that all available ecological space is already occupied (Korten, 1992) so that 'wealth creation' is the enclosure and transfer of commonly held resources into private hands. The enclosure perspective provides an 'injustice frame' (Gamson, 1995) for sections of the ecology movement including the anti-GM movement (*The Ecologist*, 1990; 1992; 1993).

The enclosure of seeds through patenting is slightly more complex than the enclosure of land. Social and environmental damage does not occur in the removal of local plants, but only at the point that new patented varieties are re-introduced displacing the local 'landraces'. Nevertheless the exclusive right of intellectual property over new varieties derived in part from seeds taken without compensation is itself an injustice, particularly given the wealth of the donors and the poverty of the recipients. The point of the enclosure argument is that, as well as any moral repugnance that may be felt over ownership of life processes, the enclosed resource generates a rental income for its new owners, often paid by its old owners. In the case of patenting rent takes the form of royalty payments. A key point here is that very little of the natural world does not belong to someone in some way. Quite frequently this ownership takes forms of common access or claim on resources. The expropriation of the local users by dominant groups claiming monopoly ownership is the routine of colonialism and capitalism (Marx, 1976).

Enclosure also destroys the accumulated local knowledge about managing the particular commons, and replaces it with disembedded expert knowledge, linked to the economic interests of the enclosers. The critique of 'development' as enclosure connects to a reappraisal of the ways in which the people of the South use the plants and animals they find in their environment (Hecht and Cockburn, 1989; Shiva, 1991). This in turn demands a reconsideration of local and indigenous knowledges, which puts in question the monopoly of knowledge claimed by western scientists, for example in plant breeding. It has been suggested that the peer review of plant breeders' work should cast a far wider net to include the local users of seeds. Scientists often use the single criterion of high yield, when assessing varieties, whereas local users often have a wider range of priorities including taste, ease of grinding and cooking (Pimbert, 1994). The struggle over the control of the economic resources of crop genetics is simultaneously a conflict between knowledge claims in different cultural projects. Local situated knowledges struggle to survive against the disembedded expert knowledges of the seed industry and biotechnology companies. Farmers make conscious decisions to breed of cultivars that possess certain characteristics, such as resistance to particular pests. Patenting genes isolated from farmers' seed is primarily an enclosure of this accumulated knowledge. It is not the process of disembedding local knowledge, but the re-embedding of expert genetic knowledge in the form of new commercial varieties that erodes existing crop biodiversity.

Any move to enclose requires an expert system for the measurement, classification, and control of property. Hence the development of various types of mapping from geometry and trigonometry for measuring taxable land areas in ancient systems, via medieval cartography (Harvey, 1993) to colonial land grabbing and botanical taxonomy (Hecht and Cockburn, 1989) and the current gene mapping. In addition to the expert systems claiming a monopoly of knowledge of these successive forms of property, there is another expert system concerning itself directly with the legitimacy of property rights as such. This legal system can draw on other expert discourses in philosophy or economics for legitimation. Expert systems themselves must be viewed as the enclosure of knowledges, legitimating only the knowledge claims that fall within their own boundaries, while delegimating any competing claims arising from the lay population.

The power of this critical discourse is that it resonates with wider anxieties typical of the age. Nuclear power acted as a lens focusing the intensity of public anxieties about centralized technocratic state control over expanding areas of daily life during the Fordist period, with big complex systems running out of human control. Genetic patenting acts as a conduit for fears about the neo-liberal project of pushing markets, private property and commodification into every part of human life. The incursions of global companies into ownership of human genes, in some ways the most intimate imaginable entity, presents a contemporary limit case of anxieties about modernity and its futures. Ownership of the genetic heritage of tomatoes or rice causes less visceral affect. Yet wealthy First World companies making up rules that allow them to claim exclusive ownership of traditional food crops of poor farmers in the South is another graphic image of neo-colonial greed and global injustice.

Alternatives to Patenting

Alternatives to patenting proposed by NGOs have taken two basic forms. The first proposal is to limit the scope of patentability. The second more recent approach is to attempt to channel the benefits of patenting towards traditional owners by assigning IPRs to them instead or as well as to the companies. Hence the concept of Farmers Rights has in the past meant the right of farmers to re-sow the seed they grow on their farms without payment of royalties irrespective of the origin of this seed. The newer notion attempts to reward farmers for their contribution towards plant breeding through supplying plant breeders with landraces with useful genetic characteristics. This latter strategy is in my opinion fraught with difficulties to the point where it becomes unworkable rhetoric. The most obvious problem is who gets paid (Yamin, 1993). Should it be the farmer in whose field a specimen is collected or the whole community of farmers who grow the same variety? What about other communities, perhaps in different countries, who also use the plant for similar purposes? The problems increase for deals with indigenous peoples. Frequently cash payments can have a corrosive effect on community organization. A series of bio-prospecting deals have been struck between companies and indigenous communities (Reid et al, 1993). Nation states have also been proposed as the suitable recipient (Kloppenburg, 1988b) of royalties. So too has the UN Food and Agriculture Organization, which would then redistribute the benefit through funding local seed conservation programmes. Any notion of community IPRs suffers from the theoretical weakness that the concept of 'rights' is defined within a modern Western discourse on individual legal subjects. (I have noted above the ideological effects of the displacement of property rights onto companies).

Far more promising is the line of argument that the rights of one 'legal personality' are limited and come into conflict with the rights of others, embodied in policies or forms of social organization in which the first claim to rights is no longer valid. Property rights, including IPRs, are not absolute but must be negotiated in relation to other sorts of rights, customs and practices. Hence in the past the rights of plant breeders to charge royalties on the seeds they produced was limited by the countervailing right, or more accurately, the custom, of farmers to re-use seed once they had grown it out on their farms. Similarly, a patent claim should, in theory, be limited by what is already publicly known or could be reasonably deduced from what is known. A right to privacy, used in the US in defending abortion, could be applied in the case of patenting human derivatives, rather than relying on property rights (Contribution to *Patents, Genes, Butterflies*, 20-21/10/1994).

Summary

Patents are texts, which embody a discourse, with its story line of progress and benefits for all. This discourse sets up an opposition between novelty and prior art and constructs a public domain as a discursive space between texts. An elitist Promethian discourse of this sort creates certain creative agents – scientists and companies – and disallows others such as farmers or local communities. Seed patenting has thrown this inter-textual public domain into conflict with other unwritten and informal discursive spaces where non-professional publics exist, and practice their own prior arts. A critical discourse argues plausibly that the current attempts to extend patenting to life forms are the enclosure of public or communal knowledge within a set of expert systems. The extension of IPRs encloses genes as an economic resource and restructures knowledge, and therefore, culture. In the next chapter I will explore how these discourses – dominant and critical, Promethian and Green Radical - have entered the field of global governance. Genetic patenting and biotechnology are discursive elements in the discourse-coalition that have raised the environmental problem of biodiversity in the FAO, the CBD and Agenda 21. Here patenting remains a contentious issue in the elaboration of a discourse of Sustainable Development. The World Trade Organization has become a tough regime, embodying a Promethian enthusiasm for gene patenting, which was more or less immune to criticism until recently.

3 Global Governance: Hegemonic Trips, Biotech and the WTO

Global Governance, Patenting, Trade and the Environment

Global governance includes a growing web of international regimes. Regimes have become central to global environmental debates due to the rapid multiplication of environment regimes reaching over 1000 by the mid-1990s, ranging from bilateral treaties to global conventions (von Molkte, 1994). The most famous of which are those relating to Biodiversity, Climate Change, Ozone and Agenda 21. Trade regimes are perceived as having important environmental impacts, particularly the Uruguay Round of the GATT signed in 1993, leading to the much wider ranging World Trade Organization (WTO) and the Seattle Round.

A regime may be defined as an ongoing international agreement, administered by an international organization embodying a particular discourse. Regimes frequently arise from rather cosy clubs of Northern industrial states with a mutual interest in regulating a particular industry on a global scale. However, a regime may subsequently open up to encompass wider global complexities. A well known example is the shift of the International Whaling Commission (IWC) from a Northern whalers club in the late 1940s to a conservationist organization by the early 1980s, with the entry of Southern non whaling countries (Stoett, 1993).

The governance of seeds and IPRs is more complex as a series of overlapping regimes have appeared. It is not immediately obvious why an organization concerned with the regulation of world trade, such as the WTO, and its predecessor GATT, should have become the vehicle for a global regime in IPRs. One line of evolution is patent regimes running for over a century from the Paris Convention, through World Intellectual Property Organization (WIPO) to the expanding trade regimes of the GATT and the WTO. A second line, concerned with the role of seeds in agriculture and the environment, runs through FAO, UPOV and thence to the CBD, connecting up with the WTO as seeds become patentable. The dominant Promethian discourse embodied in patent law broke the constraints of national sovereignty proposed in WIPO to find a more congenial home in the WTO, and has simultaneously entered the Sustainable Development discourse of the Biodiversity Convention and

Agenda 21. The TRIPs Agreement in the WTO is a key text, to which all other global agreements on genetic or seed patenting ultimately refer.

IPR Regimes: WIPO versus the GATT

Patents moved onto the international scene with the 1883 Paris Convention for the Protection of Industrial Property. Since then this convention has been updated 6 times, most recently in 1967. The Paris Convention was a classic early regime, founded by a small club of industrial nations. The World Organization on Property Rights (WIPO) was set up in 1968 as a United Nations organization to administer the Paris Convention on Patenting and the Bern Convention on Copyright, with wider representation and greater global legitimacy. WIPO was specifically dedicated to international co-operation on IPRs and was favoured by many developing countries as the appropriate forum for such negotiations. WIPO agreed that each country should be free to determine their own level of intellectual property protection, but that they should draw no distinction between their own nationals and foreigners (Subramanian, 1990). Thus in effect, establishing a series of 'level playing fields' for international competitors, but no unified global system. Within WIPO developing countries had no incentive to go any further in adopting intellectual property regimes comparable to the North, a direction they saw as against their interests. The WIPO position remained consistent with international intellectual property agreements since the Paris Convention. A basic principle of all such agreements was the sovereign right of any country to choose its own economic system. Countries at different levels of economic development are thus able to choose the level and scope of intellectual property protection they feel is appropriate (Dhanjee and Boisson de Chazournes, 1990). However, key industries in the Northern industrial countries remained unsatisfied with the strength of IPRs available to them and their enforcement. The quest for a uniform global IPRs regime has been pursued since the 1970s. Lack of success in WIPO led to negotiations being shifted to the GATT, which was not a UN organization and so governed by quite different principles. This allowed IPRs to be linked to trade issues, despite objections from the South that IPRs is not a trade issue at all.

The Institutionalization of Free Trade

To understand the choice of the GATT as the suitable forum for establishing a global IPRs regime, we have to look at the institutional nature of the GATT and the WTO. Deliberations have taken the form of pragmatic deals on trade rather than statements of principle. The GATT was formed in 1947 with the IMF and the World Bank as a global triumvirate to administer the Bretton Woods Settlement, the economic component of the post-war American hegemony. In the post-war period the USA has been the leader of the dominant industrialized countries and globally hegemonic in as far as it has been able to present its own interests as the interests of all industrialized countries and of the whole non-communist world.

The GATT was set up as an interim organization, to be replaced by a UN style International Trade Organization. However, the US refused to ratify the treaty. So the GATT remained in place from its inception in 1947, until its recent replacement by the WTO. Membership of the GATT grew from 23 states in 1947 to 116 states in 1993. By 1999 its successor, the WTO had 134 member states, with Chinese membership approaching as a US-China bilateral trade deal cleared the way. The GATT and the WTO have retained formally democratic structures relying on negotiations between member states, yet they have always been traversed by power politics, secret deals and inequality between members at every level.

Like the WTO, the GATT was an organization that promoted free trade through multi-lateral negotiations. As such it routinely attracted praise from advocates of the economic efficiency of the free market. Such praise frequently took the form of crediting the GATT with overseeing the massive growth in international trade and the concomitant industrial development. Even critics suggested that the multi-lateral rule bound nature of the GATT was superior to naked power struggles, which seemed to be the only alternative (Watkins, 1992; Acharya, 1992). Indeed the GATT was in many ways a classic liberal institution with its claim that agreed rules establish the conditions of (market) freedom and that this freedom guarantees maximum (economic) efficiency which in turn maximizes (aggregate human) welfare. Freedom, efficiency and welfare are interpreted within a Promethian discourse that supports profits for industry against protection of the environment or sensate non-humans. This becomes clear when considering the famous case of the GATT Panel decision in favour of the Mexican drift net tuna fishery as against the US environmentalists, whose concern with its effects on the welfare and survival of dolphins was considered a distraint of trade (Cheyne, 1992). International economists see free trade as an extension of the domestic free market in conventional economics Krugman (1986). Therefore international trade is thought to create maximum efficiency because it allows each country to specialize in what it does well. In North-South terms doing what you do best means sticking to the colonial division of labour. Industrial manufacture remains in the North and raw material supplies come from the South.

Liberal Hegemony: The Limits of Free Trade

Economic non-discrimination is the central principle of liberal free trade embodied in the GATT rules. This form of non-discrimination operates as a liberal hegemony to further the interests of the strongest trading nations. Industrial countries of the North (and the US in particular) bring to the negotiations considerably more power than developing countries. They have thus always been able to play a double game aimed at freezing world trade in a neo-colonial pattern by protecting home markets, while continually seeking ever greater access to foreign markets. US GATT negotiator, Carla Hills, referred to the Uruguay Round as a crowbar to open foreign markets! Needless to say, US trade and policy circles see the US as the only honest player in a world of foreign crooks and cheats (Richardson 1986).

However, the GATT, like any market, has been framed by what lies outside of it and therefore is excluded from its discourse. In a deep sense, the power differentials between the member states have always been left out of consideration. Even at a shallower level, the exclusion of certain areas of trade from its ambit has been as significant as the inclusion of other areas. These exclusions, of course, always follow from the power differentials since they depend on the relative power of member states to set the agenda.

Agriculture, for example, until the Uruguay Round, always fell outside the GATT. Timber was covered by a separate agreement, as was textiles. Hence the separate Multi-Fibre Agreement held a ceiling on access for Southern clothing and textile exporters to Northern markets. The US in particular has brought a whole range of bilateral forms of pressure to bear on its trading 'partners'. One such example is Section 301 of the US Trade Act discussed below. Another is the absurd notion of Voluntary Export Restrictions, where countries which are 'too successful' at exporting to the US 'voluntarily' limit their exports to save the US the embarrassment of breaking GATT rules to limit imports.

Free trade has always been strictly for export as far as the industrialized countries are concerned. Each state favours free trade in the areas of economic activity where it has a competitive advantage and turns to protectionism where it feels at a disadvantage. The degree of free trade embodied in the GATT simply reflected the degree to which the negotiating parties felt able to agree. Since the industrial countries have always been far more powerful than developing countries they have by and large determined the GATT agenda. Just as there is no such thing as a non-interventionist state when it comes to domestic economies, so no such thing as free trade exists. Political power is always intrinsic to the constitution of the terms of trade just as it is to the constitution of the market. Liberal free trade can be seen as a hegemonic discourse in the Gramscian sense, in that it embodies particular interests in the general rules for international trade.

Strategic Trade Policy: Games and War

In practice the industrialized countries have never played by the rules they themselves made. American neo-mercantilist economists in the 1980s abandoned all pretence of ethical rules (metaphors of freedom and non-discrimination) and favoured instead the military and paranoid metaphors of strategy and imprisonment trade policy. Strategic environments, tells us, have the 'familiar features of games and war' (Richardson, 1986: 257). Indeed the multilateral negotiating framework of the GATT was seen as too cumbersome due to its large membership. 'Bilateral trade initiatives may appear to governments, by contrast, to be much more promising' (Richardson, 1986: 265). Similarly, global welfare arguments are abandoned in favour of national benefits. 'The object of each government is taken to be the standard criterion of social welfare - the sum of domestic consumer and producer surpluses' (Branson and Klevorick, 1986: 246). However, in looking at various sectors, American economists, Branson and Klevorick, tend to find that global interests concur with US interests. They find that the US always plays fair while foreigners cheat. Explaining pressures on US trade policy requires their invocation of the 'prisoner's dilemma'. The need to break out of the prison of cheating foreigners, who go against global interests, legitimates the use of political power to attain economic objectives.

In spite of this First World manoeuvring, international trade grew 12 fold in volume under the GATT. Perhaps more significantly for the argument that follows it has also changed in its nature. For example, the UK used to import raw materials and export manufactured goods. By the 1980s, it was importing manufactured goods and exporting crude oil (Krugman, 1986). Similarly, the Third World used to be a raw material exporter and an importer of manufactured goods. Two thirds of Third World exports were manufactured goods by 1990 (Watkins, 1992). On this more fluid international scene, technology has come to be seen as the key to international competitiveness.

The Neo-mercantilist school of economists, which favours strategic trade policy (protectionism) over free trade, is in part, a response to these changes in the world economy as well as problems within economic theory. Neo-mercantilists have belatedly discovered the imperfection of the market. A large firm is now deemed to be able to affect the market through its strategic decisions rather than simply being governed by the iron law of the market. This observation leads to the recognition of entry barriers to any industry. A firm or country that develops a technological lead may exploit it without the market necessarily being able to provide competitors capable of clearing the legal and financial barriers in order to bring prices down to correspond to the average rate of profit.

A technological lead can generate excess profits in two ways - rent and overspill. A company with a lead in a new technology can charge monopoly rent on its products over and above the average rate of profit available under free competition. Patenting new inventions can provide a way of maintaining such a lead. Furthermore a technological lead may spill over into other related areas, generating a range of technologies and products and widening the base of the monopoly rents. Heavy state spending, known as strategic targeting of research and development investment, can of course enhance these profit spirals. Biotechnology remains an area of strategic investment in the USA to maintain their technological lead. In the EU similar strategic investment is aimed at challenging the American lead.

The Uruguay Round: Framework for the WTO

GATT Rounds became increasingly ambitious. The first five agreements were little more than co-ordinated sets of bilateral agreements. Uruguay was the eighth round and operated on a scale hardly imaginable in 1947. The final GATT Agreement was signed on 15 December 1993, by 116 member states and included not only trade in goods, but also services and IPRs. Negotiations took place in 14 Negotiating Groups on Goods and one Negotiating Group on Services. The complexity of the interrelated dealing in these groups has been described as three-dimensional chess. The GATT and its aptly named successor, the World Trade Organization, can be considered as organizations of truly global reach.

Begun in 1986 at Punta Del Este, the Uruguay Round was due to complete in 1990. The then Director General of the GATT, Edward Dunkel, presented a Draft Final Act (also known as the Dunkel Draft) on 21 December 1991, as an indivisible package for member states to accept or reject in total. It consists of 26 accords of 436 pages of text, of which the TRIPs Agreement on IPRs took up only 14 pages. It was this Draft Final Act which was finally signed in Geneva on 15 December 1993.

In the Uruguay Round the GATT set itself the goal of restructuring the mode of (de)regulation of international trade and establishing a framework for the 21st century under the WTO. This involved staking out a new agenda, spreading the influence of the GATT beyond its traditional area of the trade in goods to include for the first time services, such as banking, insurance and IPRs. It is no coincidence that these industries are of major importance to the US economy. Indeed the US negotiating position was frequently prepared by transnational corporations (TNCs), either individually or collectively through lobby groups. American Express presented the US position on services. While significant lobby groups included the Intellectual Property Coalition and the Multilateral Trade Negotiation coalition 'an alliance of over 200 companies, led by American Express, Citibank and IBM - set up to "inform" government thinking' (Watkins, 1992: 37).

In choosing the GATT as the terrain for a global IPRs regime, a global organization has been identified where, to quote one activist '*campaigning has little effect because the power relations are so intense*' (Interview, 12/8/1993). It was precisely this concentration of power that made the GATT such an attractive site for an Intellectual Property regime, rather than a more traditionally acceptable forum for discussion of IPRs such as WIPO.

The TRIPs Agreement

The key text in global patent regimes is the Agreement on Trade-Related Aspects of IPRs or TRIPs as it is commonly known, which was negotiated within the Uruguay Round, placing patents in the hands of the WTO. TRIPs has been called a counterfeit term (Subramanian, 1990) since it simply replicates the meaning of Intellectual Property under a spurious claim to be involved with trade. Its utility is that it has allowed Intellectual Property to be introduced into international trade negotiations, whereas previously they had seen as completely separate issues, with no logical connection. The link that has been forged between these issues is an arbitrary one, driven by very specific economic interests.

> Implicitly the *demandeurs* in TRIPs hoped to secure higher Intellectual Property protection in exchange for concessions in areas such as textiles and agriculture ... the denial of existing market access concessions was the threat for refusal to increase Intellectual Property protection (Subramanian, 1990: 510).

Subramanian (1990) locates the driving force of the TRIPS in three Northern industries - high technology, luxury goods and entertainment, with chemical/pharmaceutical and information technology companies as the leaders. Luxury goods manufacturers were concerned with counterfeit trade and therefore trade marks; entertainment and computer software manufacturers focused on copyright; and the agro-chemical giants wished to enforce patents for pharmaceuticals and 'biotechnological inventions'. The US based Intellectual Property Coalition, consisting of 13 companies, including agro-chemical conglomerates like Pfizer, Monsanto and Du Pont, advized the US Administration on TRIPs, while in the EC Unilever, Hoecht and Ciba Geigy played a similar role (Watkins, 1992).

The five issues addressed by the TRIPs were: (1) the relevance of the basic GATT Principles, (2) the standards of Intellectual Property protection afforded, (3) the methods of enforcement, (4) the settlement of disputes between governments and (5) transitional arrangements for developing countries.

TRIPs introduced the GATT Principles of National Treatment *(*GATT, 1992, *Article 3)* and Most-Favoured-Nation Treatment *(Article 4)*. National Treatment means non-discrimination between local citizens and foreign nationals. Most-Favoured-Nation Treatment means non-discrimination between foreign nationals of one country and those of another. Another important GATT Principle, Transparency, appears in Article 63 under the dispute settlement section.

The TRIPs covers seven types of IPRs - copyright, trademarks, geographical indicators, industrial designs, patents, layout-designs of integrated circuits and undisclosed information (i.e. trade secrets). The standard of IPR protection is very high. *Article 33* states that the term of patent protection is 'a period of twenty years from the filing date'. This is longer than the 17 year period that was prevalent in many industrial countries (Wheale and McNally, 1993) and far longer than in most developing countries - for example, 7 and 14 years respectively for process and product patents in India (Dubey, 1992).

On patentability, *Article 27* states that

> patents shall be available for any inventions, whether products or processes, *in all fields of technology,* provided they are new, involve an inventive step and are capable of industrial application.

Exclusions from patentability are accepted:

> to protect *ordre public* or morality, including to protect human, animal or plant life or to avoid serious prejudice to the environment, *provided that such exclusion is not made merely because the exploitation is prohibited by domestic law.*

This wording effectively removes the right of member states to exclude categories of technology or products from patentability as a part of their national policy. Further exclusions are allowed under *Article 27*:

- diagnostic, therapeutic and surgical methods for the treatment of humans or animals;
- plants and animals other than micro-organisms, and essentially biological processes for the production of plants and animals other than non-biological and micro-biological processes.

These clauses allow biotechnology to patent any genetically modified organism, since rDNA is not considered to be an essentially biological process. Certainly such is the interpretation put on the concept of *essentially biological* in the US case law following Diamond vs. Chakrabarthy (Yamin, 1993).

The final clause in *Article 27* requires 'the protection of plant varieties either by patents or by an effective *sui generis* system or by any combination thereof.' PBRs would therefore be acceptable only if they conferred rights of equal strength to patents, which in general they don't. Establishing sui generis systems thus became a focus of concerted effort in India (Interview, 17/4/1994). However, increasing pressure from the USA and biotechnology companies was directed at using the definition of PBRs given in the UPOV agreement of 1991 as the only acceptable meaning of an 'effective sui generis system'. This is a good example of how the potential political opening in one text (the TRIPs) is closed by suturing it to the already politically closed text of a previous agreement (UPOV).

Article 31 severely restricts the possibility of compulsory licensing of patented inventions. The vast majority of patents are held by Northern companies and individuals. Since many of these patents have never been worked in the South, Southern governments prefer to have compulsory licensing available to them as a policy option. India, for example, frequently utilises this option with regard to pharmaceuticals. Under the Indian Patent Act a compulsory licence may have been granted after three years (Dubey, 1992).

Parties to the TRIPs, that is WTO member states, will be required to introduce national legislation to allow patent holders to enforce their property rights. These are to include injunctions to prevent patent infringers from trading *(Article 44)* and payment of damages to compensate the patent holder, including legal fees, 'even where the infringer did not know or had no reasonable grounds to know that he was engaged in infringing activity' *(Article 45)*. Additional remedies include the disposal of goods and the implements and materials used in their production *(Article 46)*. Furthermore criminal procedures leading to fines and imprisonment are required in cases of infringement that 'are committed wilfully and on a commercial scale', *(Article 61)*. How practical all this enforcement machinery will be when applied to millions of Indian peasants using patented seeds is something of an open question. However, the Council on Trade-Related Aspects of IPRs monitors compliance of parties to their obligations.

Disputes between governments over the operation of TRIPs are dealt with in just two articles. Article 63 simply calls for transparency, that is making laws and information available in writing. Article 64 refers disputes between states to the WTO's Integrated Dispute Settlement Understanding, which will give the US, the EU and Japan greater powers to use cross-retaliation methods. They may employ access to their large markets as a lever to force Southern governments to settle in favour of the North. Southern countries have argued that disputes should be handled by WIPO, but the North was determined to keep disputes within the WTO precisely because linking IPRs to trade gives them more power.

The Transitional Arrangements in *Articles 65 and 66* allow for a phased implementation of the TRIPs. Advanced capitalist states were expected to comply within one year. Developing countries (by which is meant the Newly Industrialized Countries) had four years to comply as did countries 'in the process of transformation from a centrally-planned into a market, free enterprize economy'. Developing countries had an additional five years to extend patent product protection to new areas of technology. The Least Developed Countries (LDCs) had ten years in all, and may request technical and financial help from the developed countries.

Taken together the various points raised in the TRIPs form an agenda for the globalization of US IPRs, that is the institutionalization of the hegemonic form of IPRs as a globally uniform system. Global seed patenting serves particular industrial interests in the North including those of the biotechnology industry. However, it remained an incomplete hegemonic project, with the success of the patenting regime due to threats and intimidation as much as establishing a consensus that the TRIPs was for the general good, as became clear as the Seattle Round got underway.

Bilateral Pressure: Special Section 301

In 1988 the US Government passed the Omnibus Trade and Competitiveness Act, whose infamous Special Section 301, empowered the US to impose trade sanctions and duties on its trading partners. Under the auspices of Special Section 301, the US Trade Representative drew up a 'watch list' of countries whose IPRs were not strong enough. Brazil, India, Thailand and China all faced US action over IPRs. These countries were then targeted by the US Administration.

China provided an interesting example. On 17th January 1992, China signed an agreement with the US to protect foreign copyrights, patents and trade secrets. The eight-month dispute had just reached the point when the US was about to impose duties on Chinese imports, when a breakthrough agreement was reached. The US trade Representative, Carla Hills, was pleased with her work and stated:

> This agreement should create significant opportunities for US firms interested in marketing high value-added products to China. Principal beneficiaries will include the pharmaceutical, entertainment, computer and agrochemical industries (Hills quoted in *World Intellectual Property Report*, 1992, 6 (2): 31).

The corollary of this story was that at the November 1993 Asian Pacific summit the new US Trade Representative, Charlene Barshevsky, once again threatened action against China for lack of Intellectual Property protection. While the Secretary of State, Warren Christopher, warned of the 'unacceptable and unsustainable' trade surplus China enjoyed with the USA (Guardian, 19/11/93). It has been suggested that it was this $12 billion deficit in US trading with China that motivated the Section 301 action in the first place (Watkins, 1992). The US commitment to the principles of 'Free Trade' and multilateral negotiations are contingent upon them furthering US interests.

Similarly, Brazil was subjected to punitive tariffs on exports to the US after the Pharmaceutical Manufacturers' Association filed a complaint under Section 301 against Brazil because of a Brazilian law limiting liability to pay royalties on pharmaceutical process patents. Brazil and China were only two examples of the many countries that have been forced to introduce new domestic legislation on IPRs. The 'watch list' included 25 countries, while those that have amended their Intellectual Property legislation include Thailand, Chile, Argentina, Venezuela, South Korea, Indonesia, Taiwan, and Mexico, as well as China and Brazil. Indeed the TRIPs was made largely uncontentious by the Section 301 actions which brought most of the South to introduce domestic IPRs laws that correspond to the terms laid out in the TRIPs in advance of the signing of the TRIPs (Subramanian, 1990). A model agreement was piloted with Sri Lanka, which used the wording 'adequate and effective protection' (Porter, 1992: 14). The same wording recurred in the TRIPs, in the Biodiversity Treaty and subsequently in both the Leipzig Declaration and the Global Plan of Action on Plant Genetic Resources agreed by the FAO in their Leipzig meeting in June 1996. India, the site of by far the most vigorous anti-TRIPs demonstrations, was rapidly isolated in arguing that its own existing domestic legislation (Patent Act of 1970) was adequate and suitable to its own needs.

Technology Transfer

A question frequently posed about the TRIPs was whether or not it would encourage technology transfer. The USA claimed that strong IPRs are essential to technology transfer, while developing countries claimed IPRs, especially when owned by foreign firms, constitute a barrier to technology transfer. The economic performance of a country is thought to depend, not on its natural endowments of resources, but on its technology. Yet technology is more than simply the hardware employed to make commodities. According to the World Bank 'Technology is the knowledge that leads to improved machinery, products, and processes... Technology also includes knowledge embodied in management know-how' (World Bank, 1991: 88). It need not even be written down. The same World Bank Report distinguishes between embodied and disembodied technology. If technology is a form of knowledge, then IPRs replace more conventional property rights as the key to controlling technology and hence wealth. It follows from this whole line of thinking that if access to technology is the source of wealth, then the level of development of a country can be judged by transfer of new technologies to it from more technologically advanced countries. Hence the centrality of technology transfer to recent development thinking (Acharya, 1992).

Given the importance accorded to technology, the USA argued that US companies would not make the technology they own available to developing countries unless they are guaranteed control of that technology through IPRs. Many developing countries, on the contrary, were prepared to take that risk and prefer to copy or licence foreign technology, thus bringing it under their own control. The US argument is clearly mendacious since it was combined with complaints that developing countries steal their technology - that is copy it. Far from encouraging technology transfer the TRIPs regime will simply give the industrialized countries in general and the US biotechnology industry in particular a way of ensuring they can collect the rent on their technological lead. 'TRIPs is more accurately seen as an exercise in rent creation and rent shifting' (Subramanian, 1990: 516). TRIPs extracts yet more profits from the poor countries of the South for the benefit of Northern industry.

Throughout the deliberations of the UNCED Rio Summit and its numerous sub-conferences technology transfer recurs as a cypher for the whole question of North-South inequality and the need for social justice. Over and over again it is included in the package to make sure that the South won't get left behind. This Promethian technicist discourse allows environmental problems to be defined as a lack of technology rather than a surfeit thereof. It also defines the issue of development as an attempt to get the South to catch up with the North, rather than a social justice issue of the distribution of finite resources. The South is then seen simply as a deficit of liberal capitalism rather than being composed of other countries with other cultures. This focus on technology transfer generates what many environmentalists would call techno-fix solutions to environmental problems. For example, Pimbert (1994) argues for the replacement of the expert-led technology transfer model of agricultural research to be replaced by one that is rooted in dialogue with the more complex vernacular knowledge of local farmers.

Sustainable Development: the Biodiversity Convention and the FAO

IPRs are not restricted to the Promethian world of the WTO, they have also penetrated the set of international agreements that embody a discourse on

Sustainable Development. The key attribute of Sustainable Development is that it attempts to find a win-win solution to the competing claims of economic growth, social equity and environmental protection. The Biotechnology and Agribusiness lobbies can be relied on to drag sustainability in the direction of Promethian growth and profits. NGOs, with their Green Radical discourse, attempt to hold these institutions to account on equity and environmental issues. Sustainable development is often the outcome of the ensuing struggle. Two regimes concerned with plant genetics that can be understood as committed to sustainable development are the Convention on Biological Diversity and the Food and Agriculture Organization (FAO).

Developing countries had previously been able to get the FAO to adopt an Undertaking in 1983, which called for all seeds to be common heritage, including inbred elite lines used for breeding by seed companies. However, industrialized countries introduced an 'Agreed Interpretation' in 1989 that the seed companies' IPRs were 'not incompatible' with the undertaking (Porter, 1992). While TRIPs awaited the agreement of the GATT package, the same arguments about IPRs versus technology transfer blew up at the Earth Summit in Rio de Janeiro in June 1992, when the Biodiversity Convention was to become enshrined in international law. The negotiations towards the Biodiversity Convention saw a change in Southern tactics, under Indian leadership. Instead of calling for 'common heritage' they stressed the recognition of the 'sovereign rights of states over their natural resources' (Porter, 1992: 5). Access to Southern genetic resources was then to be linked to access to Northern biotechnology based on Southern germplasm. The thrust of Southern arguments was to ensure technology transfer through compulsory licensing, and other limits on IPRs.

The biotechnology industry exerted considerable pressure on the Bush Administration during the negotiation process. The Industrial Biotechnology Association (IBA), representing 80% of the US Biotech companies, expressed its desire that the Biodiversity Convention be made subordinate to the provisions of the GATT (Porter, 1992). The IBA as well as the Pharmaceutical Manufacturers Association (PMA) and the Association of Biotechnology Companies wrote to Bush complaining that the draft texts did not guarantee strong enough IPRs (Porter, 1992). Unsurprisingly, Bush refused to sign the ensuing treaty, and was immediately congratulated by the IBA and PMA even though the US had been able to insert in *Article 16* the following crucial sentence:

> In the case of technology subject to patents and other intellectual property rights, such access and transfer shall be provided *on terms which recognize and are consistent with the adequate and effective protection of intellectual property rights* (UNCED, 1992, Article 16; Porter, 1992: 13, my emphasis).

An identical clause, using the exact same wording recurs in the subsequent Leipzig Declaration and Global Plan of Action on Plant Genetic Resources to come out of the FAO deliberations in June 1996. The mantra of recognition of adequate and effective IPRs can be expected to re-appear as a reference point in all future global negotiations on the subject. Porter goes on to argue that the wording 'adequate and effective protection' means the Biodiversity Treaty would be governed by the TRIPs and US bilateral agreements discussed above, and so that the US has nothing to fear from the Treaty. In this he is explicitly rejecting the interpretation advocated by India (and feared by Bush), which claims that compulsory licensing is still an option.

The Clinton Administration subsequently agreed to sign on the basis of the interpretation put forward by Porter in his position paper. 'As all diplomats, international lawyers and civil servants know, "Interpretation is everything"' (Cameron and Ward, 1992). The decision to sign was taken after talks with the World Resources Institute (WRI) and two other environmental groups in conjunction with the Merck Pharmaceutical Company plus a further two TNCs (Mooney, 1993). On the basis of the Porter interpretation, these American NGOs reconciled the Biodiversity Treaty with TRIPs, and so overcame the isolation of the USA. The project to give the poor but gene rich South a fair deal was subordinated to the hegemonic project driven by the US biotechnology industry. It is precisely through articulating new initiatives such as biodiversity, through dominant discourses of free trade, genetics and patenting, that biotechnology assumes its hegemonic status.

The strategic play by WRI and their allies demonstrated very clearly two dimension of what I have called the counter-expert role of NGOs. On the one hand they were able to supply a very skilful (i.e. expert) legal interpretation required to get the USA to sign the Biodiversity Convention. On the other they were able to mobilise the political legitimacy attached to environmentalism in the service of US policy making. However, this act of 'legitimacy broking' was extremely contentious, sharply dividing the 'NGO community' as many saw WRI as 'gutting the Biodiversity Treaty' (Interview, 31/3/1994).

Extending the WTO Agenda: The Seattle Round

Gene patenting occupied a central place on the proposed agenda for the launch of the millennial round of WTO negotiations at the ministerial meeting in Seattle in December, 1999. Proposals came from two opposing points of view.

> The first was the US call for new rights for companies to trade genetically modified organisms without trade restrictions....
> A key demand for developing countries was to stop the WTO's patent rules from allowing foreign companies to rip off their biological resources (Coates, 1999: 4-5).

The Miami Group of GMO exporting countries - USA, Canada, Argentina, Uruguay, Chile and Australia - who had come together to oppose the Biosafety Protocol, wished to prevent controls on importation of GMOs by the EU and other countries, and argued that they were a distraint of trade. Proposals from the USA, Canada and Japan aimed at establishing a science-based WTO Working Group on Biotechnology to smooth the path of international trade in GM food, were included in the draft Seattle Ministerial Text (CIEL, 1999).

Kenya, on behalf of the Africa Group, and supported by the NGOs, proposed a review of article 27.3b of the TRIPs Agreement.

> The review process should clarify that plants and animals as well as micro-organisms and all other living organisms cannot be patented, and that natural processes that produce plants, animals and other living organisms should also not be patentable (WT/GC/W/302, quoted in Third World Network, 1999).

The Kenyan proposal argued that these were life patents and therefore contravened patent law. Furthermore they pointed out that the provision for a *sui generis* equivalent to patenting should

> protect innovations of indigenous and local farming communities ... allow the continuation of traditional farming practices including the right to save and exchange seeds and sell their harvests ... prevent anti-competitive rights or practices that threaten food sovereignty of people in developing countries (Third World Network, 1999).

The Ministerial Meeting in Seattle ended in a shambles, without agreeing an agenda. The negotiating process was widely seen as much of a stumbling block as the substantial policy issues on the table. Corporate representatives gathered to socialize with the WTO and heads of state. One such representative whose corporation paid $250 000 for access to meetings and receptions, denied any attempt to influence the WTO. 'It's not buying influence. We don't need to. We are in daily contact with the WTO and government' (quoted in Vidal, 1999a: 15). Developing countries were excluded from decision making from the start. Only 20 of the 135 member states were invited to the prior negotiations in the WTO's 'Green Room' in Geneva (Atkinson and Denny, 1999). In Seattle the real talks continued behind closed doors ('The Americans are in hotels sewing up deals') (Vidal, 1999b: 1). In Seattle the WTO for the first time became the centre of public interest with mass protests outside the conference centre disrupting the proceedings and continuous media coverage. Emboldened by the mass protests outside, the developing countries led by the African Group refused to accept US proposals (Vidal and Elliot, 1999: 2). EU negotiators conceded to the USA and the Miami Group on trade in GM food, but individual EU ministers rejected this concession. The director general of the WTO, Mike Moore, was criticised for lack of leadership in the logjam. 'There is general dissatisfaction that Mr Moore has not circulated his own text. The view was that the director general had a responsibility to give a bit of a lead' (EU scurce quoted in Elliot, 1999). As the WTO policy-making process faced a crisis of legitimacy, exposed by a massive intervention from global civil society, the hegemonic project was once more put into question.

Summary

The signing of the GATT Uruguay Round in December 1993 was a nodal event in the history of the global regulation of seeds. The Uruguay Round set up the WTO and included the TRIPs, which legalized seed patenting globally. The TRIPs restructured power relations in the field and became a point of reference for subsequent international agreements concerning intellectual property over seeds and hence the global extension and integration of the enclosing expert systems described in the previous chapter. The attempt to establish uniform global IPRs over living material was a hegemonic project driven by the biotechnology industry with its complex articulations of expert systems and economic interests. Hegemony needs to be understood through a blend of Gramscian and regime theory as an attempt to contingently construct particular interests into an apparently universal good. The GATT provided a fertile terrain for the hegemonic ambitions of the biotechnology industry, even though intellectual property had been a recent import into the world trade regime.

The emerging biotechnology industry in the US used its influence to extend the limits of private property so as to include life forms. An impressive range of social actors and discourses were articulated into a discourse coalition around a hegemonic project. TNCs based in pharmaceuticals and agro-chemicals enrolled not only their lobbyists in Washington, Brussels and Geneva, but also political actors in numerous departments and committees in the USA and the EU. At a deeper level this hegemonic project depended on the articulation of expert discourses of the dominant paradigms of reductive technoscientific endeavour from molecular biology, through trade economics to patent law. This allowed the global enclosure of genes to appear as the natural outcome of technoscientific progress and the institutional logic of free trade rather than the economic interests of particular companies.

The GATT and WTO provided a suitable setting for this hegemonic ambition. The liberal free trade regime of the GATT embodied concentrated economic power within a universalising discourse on trade. The Uruguay Round with its tilt towards neo-mercantilism further intensified the power relations brought to bear on the implementation of a new intellectual property regime, which was achieved by bilateral pressure as much as multilateral negotiations. This outcome was not the inevitable outcome of economic progress, but the imposition of a single viewpoint and the suppression of feasible alternatives. There has, however, been resistance to the appearance of biotechnology and this global drive for gene patenting, both in the resistance of many countries to signing the TRIPs and in the appearance of an oppositional movement, as has become increasingly apparent.

The application of patent law to genes and seeds combined the extension of an expert discourse with the enclosure of commonly held socio-ecological resources. The extension of patent law through the TRIPs embodied a global hegemonic project, in which the economic interests of the biotechnology industry, particularly the agro-chemical transnationals, were framed in an expert discourse of patent law. The TRIPs was a foregone conclusion in that the USA had forced most opponents of TRIPs to include equivalently strict IPRs into their domestic law by threatening them with Section 301 sanctions. Furthermore the possibility of the phasing out of textile quotas in the Uruguay Round alone was enough to encourage unwilling Southern countries to deal. The TRIPs was not a free trade policy. Nor was it one that was in the best interests of the world as a whole. TRIPs was introduced as a means of siphoning resources from the poor countries of the South to the rich ones of the North.

The distribution of power and legitimacy between international organizations is uneven and the WTO fell on the side of the deployment of power to regulate, while other fora such as the FAO and CBD delivered legitimacy. The classical Greek distinction between 'hegemonia' (leadership) and 'arkhe' (the power to rule) is instructive (Perlman, 1991). Hegemony is a form of legitimate domination, hence the importance of establishing TRIPs as an institutional master frame, and legal precedent. A global hegemonic project needs to be re-embedded in popular consciousness in at least some key localities in order to maintain legitimacy. The USA has sometimes been isolated in some of the 'softer' international forums, and more recently even in the WTO. Such crises of legitimacy revealed a certain weakness in the hegemonic project, which threatened to descend into the exercise of naked power.

4 The Counter-Expert Challenge: NGO Leadership in the Anti-GM Movement

Counter-Experts in Context

A global anti-GM movement, embodying a Green Radical discourse, opposed the global hegemonic project of seed patenting. The leadership of this movement, at least until the mid-1990s, took the form of individual counter-experts setting up and working in Non Governmental Organizations (NGOs). This chapter is based on interviews conducted with movement leaders in the UK. This account needs to be set in the context of three aspects of social movements. Social movements consist of linked social networks. They frame issues and identities discursively; and they develop changing repertoires of collective action through their lifecycles to achieve their aims. These aspects are discussed in more detail, followed by accounts of key movement leaders and their organizations.

Mapping the NGO Networks

The social ecology of the 'NGO Community' that formed the leadership of the anti-GM movement in the UK in the mid-1990s contained several niches occupied by different individuals and NGOs. While there was a larger group of NGOs that had developed an interest in the topics of biotechnology and biodiversity, this chapter concentrates on the key actors who shaped the critical discourse on biotechnology and patenting in the UK.

The NGOs presented in this chapter are summarized in table 1. Accounts of the NGOs follow the order of their appearance in the table. Each NGO had limited available resources of time, money, membership and expertise. Each actor had his/her own perceptions of the movement networks and contributed to constructing these networks. Each actor framed the issues to be addressed, drew limits to these issues, and prioritised within that range of issues, influenced by their own experience and preferences. The NGOs each had to decide whether to engage local, national, European or global governance, although the cycles of activity by their adversaries forced the NGOs to regularly shift their emphasis from one political arena to another.

Table 1: Seed NGOs

Name	Membership, (Foundation Date)	Key Interest	Primary Network	Main Discursive Frame	Policy Level
Green Alliance	380 (1978)	GMO Releases	Anti-Biotech	Environment Risk	UK
Genetics Forum	31 (1989)	Genetic Engineering	Anti-Biotech	Human and Environment Risk	UK, Europe
Greenpeace UK	400,000 (1977)	GMO Releases, Patents	Anti-Biotech	Environment Risk	Global
ICDA, RAFI, GRAIN	(1979)	Farmers, Seeds, IPRs	Seed	Rights	Global Regimes
Farmers' Link	300 (1986)	Sustainable Agriculture	Seed, Anti-Biotech	Environment Risk	Local to Global
Intermediate Technology	Non-member (1965)	In-situ Seed Conservation	Seed	Rights	Global Regimes
Gaia Foundation	Non-member (1984)	Indigenous People	Seed	Rights	Third World Network
WWF International	8000000 (1961)	Nature Conservation	Seed	Environment Risks	Global
Heritage Seed Library	9,000 (1974)	Seed Saving	Seed	Environment Risks	UK

Any typology of the NGOs involved in the anti-GM movement needs to Emphasise a central divide on the scale of the organizations. On the one hand were the small specialist information-rich NGOs with a handful of workers and memberships of less than a thousand. Examples of this group of NGOs were Genetics Forum, Green Alliance, GRAIN, RAFI, the Gaia Foundation and Farmers' Link. On the other hand were the large membership organizations, Greenpeace and World Wide Fund for Nature (WWF). These were usually named as the only two global NGOs interested in the area of genetic engineering and patenting. Even in such large NGOs the presence of individual counter-experts are of great importance for the organization's participation in the anti-GM movement. Falling between these two extremes is Intermediate Technology (IT), which has offices in six countries and a role in direct promotion of seed saving in the South, but also acts in the UK as a counter-expert NGO, without a membership base.

Two broad types of network overlapped among the NGOs: anti-Biotech and seeds. The originator of the anti-Biotech networks was Jeremy Rifkin in the USA. These networks, which stressed the risks associated with biotechnology as a whole, extended to the UK in 1987 through Julie Hill in the Green Alliance and the formation of the Genetics Forum. Greenpeace subsequently became an actor but, unlike Genetics Forum, remained marginal to the European network, GENET.

The originator of the seed networks was Pat Mooney with the ICDA Seeds Campaign in Europe in 1979, which later divided into GRAIN and RAFI. The seed network originated within a development NGO milieu, and placed a strong emphasis on the rights of farmers (peasants) to equity and autonomy. Pat Mooney recruited Patrick Mulvany of IT, who set up the first seed conference in the UK in 1992, and the Food for Life network from 1995 onwards.

Each NGO was at the centre of its own network. While there were numerous connections, 3 nodal points appeared in the UK NGO community: Genetics Forum as co-ordinator of the anti-biotech groups, IT as co-ordinator of global seed campaigns and HDRA as the centre of domestic seed saving. Gaia was able and willing to intervene in both anti-biotech and seed networks. Liz Hosken (Gaia) shared a central role in the global leadership at the gathering of NGOs at the FAO meeting in Leipzig in 1996 with Pat Mooney (RAFI), Henk Hobbelink (GRAIN) and Vandana Shiva.

The preferred level of operation of the NGOs appeared to range from the truly global Greenpeace and WWF; through the internationally linked IT, Gaia and GRAIN; and the national Genetics Forum and Green Alliance; to the resolutely local Farmers' Link, with its base in Norfolk farming communities. However, all of these NGOs attended the Earth Summit in Rio. International connections were part of the everyday life of NGOs. RAFI and GRAIN in particular had global reputations and were angled primarily to influencing global negotiations. One way or another, all of them participated in global civil society.

Anti-GM Discursive Frames

A major role for social movement leaders is to develop a credible discursive story line, which frames the conflict they are engaged in and the adversaries they confront in a way their followers can identify with. A composite anti-GM frame has evolved from a social justice critique of the development paradigm of the Green Revolution, hybridised with the 'global biodiversity crisis' discourse in nature conservation circles, as well as the technology risk discourse on biotechnology. Different versions of the anti-GM frame are made up of combinations of discourses on risks and rights, summarised in table 2.

Table 2: Anti-GM Frames

Risks	Health (Food Quality)
	Environment (Biodiversity Loss)
Rights	Autonomy (Control over Seeds)
	Equity (Compensation)

Thus NGOs have developed two major discursive frames that organised the conflict over biotechnology: firstly an explicitly environmental discourse of technological risks; and secondly a discourse of human rights and equitable distribution of resources, that fits a more traditional pattern of an 'injustice frame' (Gamson, 1995). These discursive frames corresponded to the two aspects of patenting - embodiments of technological knowledge and property rights. The risk frame is primarily a critique of the impacts of biotechnology as new knowledge applied to humans and the environment, that is, as a technology. Leaving aside arguments about eugenic the application of biotechnology directly to human reproduction, biotechnology could pose a direct threat to human health through its impact on food quality. With the subsequent marketing of GM food, food quality has sharpened as a sub-frame. Biotechnology may also have an impact on the environment, leading to biodiversity loss, with indirect effects back on humans. A moderate version of the environmental risk frame focused on tighter control of environmental releases of GMOs. A stronger version of the environmental risk frame rejected biotechnology *in toto* as a risk to the integrity and intrinsic value of non-human species, breaking fixed natural boundaries between these species.

The second discursive frame was a more familiar justice frame, based on an appeal to human rights. It blended two ethical positions (a) the autonomy of individuals, communities and even nations and (b) equity in the distribution of resources between companies and communities, or between North and South. The rights frame was deployed against patents as property rights. The autonomy version asserted the right of farmers to have free access to seeds, to save seed, breed and re-plant, without paying royalties to anyone else. That is that they should not be simply means to others ends. (Nature is also sometimes seen as having its autonomy compromised by genetic patenting, which reduces nature to simply a means to human ends through commercialization). The equity version attacked the negative distributional effects of patenting as 'biopiracy' and enclosure of genetic and knowledge resources. This led to compensation claims against biotechnology companies, on behalf of farmers, for their *in-situ* conservation of seeds.

These frames did not automatically deliver judgements on what was just or sustainable. Within the process of rhetorical dispute actors creatively weave together strands of separate, and even conflicting, discourses (Billig, 1995) to make novel judgements, not governed by pre-existing criteria (Lyotard, 1988). What risk are acceptable or which rights farmers should have, was the subject of novel claims which rested on analogy to more established discourses. Thus each NGO blended risks and rights discourses to frame the issues in its own way. For example, Genetics Forum was set up as an anti-Biotech coalition, and included a range of viewpoints, including Greenpeace and Green Alliance, who concentrated on the environmental risks of GMO releases, with Greenpeace taking more of a hard line than Green Alliance. GRAIN, RAFI, IT and Farmers' Link foregrounded agricultural biodiversity and linked it to farmers' rights and food security. Gaia linked a strong version of both risk and rights frames to the indigenous peoples' movement and thus a certain level of Southern spiritual critique. WWF was divided between a dominant group for whom nature conservation is the only concern, and a small group concerned with agricultural biodiversity. The NGOs spent a considerable amount of energy on 'frame alignment' (Snow and Benford, 1988) in the process of assembling various coalitions or networks.

Movement Lifecycles and Action Repertoires

Over time social movements follow their own lifecycles, with different types of actors coming to the fore in the various stages of their life cycle. Much attention is given to periods in which social movements engage in protest or direct action and expand the available repertoire of contentious collective action (Tarrow, 1994). However, this 'mobilization' forms only the tip of the iceberg of social movements, with the much larger and more continuous 'latent' networks hidden from public view (Melucci, 1989). In earlier and later stages less disruptive forms of action are likely to predominate. This chapter is concerned with an early stage of the anti-GM movement, in which a small counter-expert leadership was coalescing into a public interest lobby, commonly called the NGO Community in the UK, and developing the arguments that would later feed a wider movement.

Table 3: Environmental Organizations

	Conventional	Disruptive
Professionalxc	Public Interest Lobbies	Professional Protest
Participatory	Mass Advocacy	Mass Protest

Environmental organizations have been classified according to their degree of professionalization and type of collective action repertoire, as in table 3 (based on Diani, 1997). Much interest and concern has been voiced over the shift over time from the bottom right quadrant to the top left: from mass protest the professional lobbying in the later stages of social movement cycles (Diani, 1997; Jamison, 1996). Yet in earlier stages the reverse is true, especially where high technology is at stake. While this schema has certain weaknesses, namely that it does not allow for innovative forms of non-confrontational action, it is useful to apply Diani's typology to the anti-GM movement, as shown in table 4. All the organizations in this chapter fall into the top left quadrant. (For those in the bottom left quadrant see Purdue, 2000).

Table 4: The Anti-GM Movement

	Conventional	Disruptive
Professional	NGOs	Greenpeace
Participatory	Seed Savers	Anti-GM Protestors

Jamison and Eyerman (1991) view social movements in terms of innovation as well as confrontation. They have highlighted the importance of intellectuals, or dissident experts, in the early stages of social movements. As these intellectuals interact within a movement culture they produce two categories of cognitive praxis: counter-experts that challenge expert formulations of risk and regulation and 'grass roots engineers' who develop alternative technologies. The notion of 'iconic praxis' (Harris, 1994) has been used to signify the media-friendly symbolic action of Greenpeace, who have proprietary rights over the top right hand quadrant in all these diagrams. Seed savers are hidden enclaves of cultural innovation. Counter-experts have an important relationship with green enclaves in that they both frame environmental issues (Gamson, 1995) and claim to represent these networks in institutional fora, thus challenging the legitimacy of the state and expert systems as representative of public feeling. These alternative modes of mass participation in social movements during latent and mobilization phases have been described as 'enclave building' and 'legitimation stripping' (Welsh, 1988). Legitimation stripping is a disruptive collective action repertoire that focuses media attention on the bodies of the protestors as subjects of discipline and punishment (Foucault, 1977). Authorities and agencies of social control find themselves unable to rely on internalized social discipline to control the protestors and thus move to physical punishment, for example - dragging protestors away, arresting them, imprisoning them or using tear gas, batons and rubber bullets. Legitimation stripping signals the collapse of the legitimacy of the dominant forces in the eyes of the protestors, indicated by the breakdown of internalized social discipline in the face of the symbols of authority. The use of violence by the police towards protestors often attracts the media, which translates the process of legitimation stripping back from physical violence to symbolic contest. A wider public then becomes exposed to the spectacle of political authorities acting as if they lacked legitimate consent from civil society and were able only to continue in power through direct force. Thus the purpose of direct action is to separate the power of the state from its legitimacy rooted in civil society.

Table 5: Collective Action Repertoires

	Latent Innovation	Bodily Confrontation
Professional	Counter-Expert Framing	Iconic Praxis
Participatory	Enclave Building	Legitimation Stripping

These categories are intended only as a guide, counter-experts are an affectual community as well as a cognitive one, as are seed savers. Networks, frames and action repertoires develop simultaneously, interacting with each other, and all have a spatial dimension. The movement leaders, even at an early stage in the development of the anti-GM movement were concerned with developing global networks, global frames and an action repertoire designed to engage with global governance. Action repertoires are not purely mechanical combinations of tactics. Social movements have to assess different ways of taking action and convince themselves of its effectiveness. An important part of collective action is developing a sense that the movement can achieve the kinds of change they wish to see. Developing an appropriate action repertoire goes hand in hand with evolving the agency of the movement (Gamson, 1995). Movement leaders play a significant role in network building, framing key issues and proposing action and developing the agency of the movement.

Movement Leadership

Leaders may be viewed primarily in terms of their ability to transform their situation or in terms of the pattern of their transactions with their followers, or in terms of the way in which their leadership style is contingent upon the environment they work in. Theories of transformational leadership (Bryman, 1992; Moscovici, 1993; 1994) use Weber's concept of charisma to focus on the ability of the leader, to generate trusting followers. Theories of transactional leadership (Hollander, 1993; Melucci, 1996) give a much stronger role to followers, in that leaders and followers engage in mutually dependent symbolic exchanges (Simmel, 1955; 1971; 1986; Goffman, 1969; Hannerz, 1980; Hollander, 1993). Theories of contingency (Bryman, 1992; Chemers, 1993) stress the situation in which leadership takes place, generating a range of leadership styles.

Identifying leadership within social movements and social networks relies less on locating who holds formal positions than on the reputation of individuals among their peers. Unlike political parties, social movements lack formal structures and deny formal representational and decision-making roles for leaders resulting in a form of concealed leadership, where interested individuals take a lead on issues that concern them (Melucci, 1996). Movement leaders are engaged in network building and occupy key points in flows of information. One key point is as a reference person at the centre of the whole network or a cluster within the network. A second key point may be a bridge point between two clusters or networks, where the broker sits. The broker's power lies in making and controlling connections between networks (Hannerz, 1980). Key leadership resources, of particular relevance to counter-experts, are the discursive frames that define group identities and purposes. 'Leaders define situations for followers. Yet followers must be willing to buy that definition' (Hollander, 1993: 32). Thus leaders must propose objectives and the means of achieving them (Melucci, 1996) that are convincing to their potential followers (Snow and Benford, 1988).

Anti-GM NGOs in the UK

Green Alliance: Releases, Risks and Reflexivity

Green Alliance was set up by a group of eminent people in 1978 who felt that as there was no chance of a Green electoral victory, an NGO was needed to 'Green' the established parties and later industry and Whitehall. The residue of this original alliance remained in a group of about 400 patrons, with a more active board of 8 people. The Green Alliance employed 5 workers of which only the director, Julie Hill, worked on biotechnology with part time support from one assistant. The Green Alliance's interest in biotechnology emerged as part of a strategy of identifying new green issues and trying to get them on the political agenda. Their continued work on the environmental release of GMOs reflected a lack of activity by the major NGOs whom they had hoped would have taken over the whole issue of biotechnology much earlier. In fact *'none of the organizations have taken the decision to make it a major campaigning issue'* (Interview, 5/5/95). Biotechnology was the only specific issue they took up within their broader remit of generic issues such as environmental taxation or monitoring the 'Greening Government' initiative. *'We don't do ozone, or wildlife or water or whatever'* (Interview, 5/5/95).

The Green Alliance's work on biotechnology and GMO releases started with working on the visit to the UK of the American campaigner, Jeremy

Rifkin in 1987. Rifkin returned to the US and continues to be a major player there but his assistant, Linda Bullard, remained in Europe, working first in Germany and then in Brussels where she advized the Green Group of MEPs in the European Parliament. In the UK, Julie Hill and her then colleague at the Green Alliance, Nick Rowecliffe, together with Hannah Pearce and Eric Brenner set up the umbrella group, Genetics Forum.

Julie Hill sat on ACRE; the government committee that oversees the release of genetically modified organisms into the environment. She drew a clear distinction between the issues of release and patenting along two axes, ethical versus environmental and national versus international.

> *I sort of decided quite early on in the patenting debate, that while I think it is a very important set of probably predominantly ethical and moral issues, and practical issues for other countries, its not about environmental safety, certainly in the UK. So I decided to stay out of it, because we're an environmental group* (Interview, 5/5/95).

This separation depended on limited resources of time and expertise. It was also a specific strategic choice to work at a national level within the UK. At the same time she chose to work on issues that could be much more closely defined by their impact on the environment. She contrasted this approach with Greenpeace.

> *A group like Greenpeace, where they're looking at things very holistically, take into account all the ethics, and human dimensions of the industry developments as well as environmental ones* (Interview, 5/5/95).

She supplemented these clear distinctions by pointing out that patenting could have indirect environmental implications affecting what comes to market and therefore what gets released. It would also affect seed saving.

> *The saved seed could be a very big issue for small farmers, or small farming systems, but that's not something we've got into* (Interview, 5/5/95).

Issues may and do get linked in quite different ways. Lyotard has likened the construction of a discourse to commerce between islands: any number of routes may be chosen to connect these islands. Consequently, any political project requires judgements to be made (Lyotard, 1988). Julie Hill made her judgements explicit in narrating the route she had chosen to carve out a clearly defined niche for herself and her organization.

Whether risks to agricultural biodiversity can be seen as environmental impacts or not depends on the social construction of 'the environment'. A substantial part of the work of the ACRE committee consisted in defining what counted as 'the environment' for the purposes of policing the boundaries of the committee's responsibilities[6]. A key distinction was made between the 'natural environment', which fell into their remit, and agriculture which did not.

The Green Alliance's primary focus on the UK did not exclude some international work, at Rio and elsewhere, on developing a Bio-Safety Protocol. While biodiversity as a whole fell outside Julie Hill's brief, the Bio-Safety Protocol which had been attached to the Biodiversity Convention, allowed her to cross over, but her focus remained on the safety of GMO releases. As far as she was concerned the dispute followed predictable lines.

> *It's the usual thing, the G77 countries say they would like one and the Americans and the Japanese say there mustn't be one. It depends entirely on the politics how that plays out* (Interview, 5/5/95).

She considered Greenpeace International to be the major NGO player, which took part in the Bio-Safety negotiations.

> *They're pretty well involved in the international regime politics, which is a set of expertise all of its own. How to do international negotiations and talk about law is completely different to worrying about release* (Interview, 5/5/95).

In Europe she was involved with discussions over regulations and de-regulation. This in turn depended on getting information through an established network, which extended to Europe, where Linda Bullard was still in contact and acted as an important information source on European politics. The Co-ordination Europeenne Des Amis de la Terre (CEAT) biotechnology clearing house based in Brussels' Friends of the Earth was useful too, but a lot depended on the co-ordinator.

> *A lot of these things are really down to personality. It really depends on if the person involved knows what's going on, and knows lots of people, but the paper work is not fantastically useful* (Interview, 5/5/95).

Within the UK the Green Alliance was *'at the centre of an extremely large network, or we are part of an extremely large network'* (Interview, 5/5/95). They put together a loose coalition on biotechnology to lobby the Department of the Environment - including Greenpeace, WWF, National

---

6 The role of the chairperson, Bristol University's John Beringer, in this boundary work was highly significant. For the politics of boundary work in science see Haraway (1992); Gieryn (1995).

Council of Women, Farmers' Link and Genetics Forum. The Green Alliance serviced this coalition - setting up the meetings, taking the minutes and keeping all the NGOs in touch with each other. The coalition contained a mix of NGOs. On the one hand, there were the generalist mass membership organizations, such as Greenpeace, WWF and the National Council of Women, which were *'such large bodies they do tend to slightly frighten civil servants'* (Interview, 5/5/95). On the other hand, were the small, specialist, information-dense NGOs such as Genetics Forum, Farmers' Link and the Green Alliance itself. In combination these two forms of NGO provided a blend of counter-expertise and constituency support among the public. A mass membership provides an NGO with a basis for fundraising as well as political weight. Small NGOs must rely on the quality of their arguments for effective lobbying and on their output of publications for fundraising. *'The product is very important in fundraising for small NGOs'* (Interview, 5/5/95).

Genetics Forum: Coalition Politics

Genetics Forum was heavily cited by other members of my snowball sample as the UK umbrella group co-ordinating campaigning against genetic engineering. As such, it expressed most explicitly a key dimension of the NGO networks: the formation of coalitions around points of agreement.

> *A Coalition involves everyone agreeing to work on issues they all agree on, and not to publicly undermine the coalition, while continuing to work separately on issues they can't agree on.* (Interview, 27/6/95).

This focus was reflected both in the nature of its membership and its campaign structure. Thus Genetics Forum had only 31 actual members; drawn by invitation from other NGOs, though individual members of the public subscribed to its newsletter, *Splice of Life*. Genetics Forum organized a series of NGO coalitions on the release, patenting and marketing of GMOs, with a further coalition to oppose the use of the genetically engineered bovine growth hormone, rBST, in dairy farming. Genetics Forum also represented the UK in GENET, the European network against genetic engineering.

Genetics Forum's first incarnation employed David King, a geneticist. He was not re-employed when Genetics Forum was re-activated in 1994, after a period of inactivity. He then set up a separate news service, *GenEthics News*, but remained a member of Genetics Forum able to attend the AGM and question the direction of the organization (AGM, 28/9/95). It also meant that in a very understaffed area there were two publications, *GenEthics News* and

Genetics Forum's own *Splice of Life*, covering the same issues with no major differences of analysis.

During the period of my research Genetics Forum was run by two part-time employees, Robin Jenkins and Steve Emmott. Steve Emmott subsequently moved to a job at the CEAT Clearing House on Biotechnology in Brussels and Robin Jenkins started farming in France, but continued working for Genetics Forum, and Julie Shepherd was brought in to produce *Splice of Life*. Robin Jenkins had a lot of prior experience in food politics including working for the GLC Food Policy and ILEA school meals sections. Robin dated his interest in genetic resources to the *Gene Traders Conference* organized by IT in 1992. Julie Shepherd too was recruited from the SAFE Alliance, and therefore also had particular links with the healthy food lobby. A large percentage of their budget was derived from the Goldsmith Trust on a fairly open brief to campaign against biotechnology in the food industry (AGM, 28/9/95). Thus the emphasis of their funding and expertise shifted in the direction of GM food.

Six-monthly meetings of their eight-person management committee monitored the employees, with the AGM as the opportunity for participation by all members (and interested outsiders). The 1995 AGM displayed a rather defensive control of the agenda by the chair and secretary resisting dissidence from David King and Les Levidow. It was not clear how much this was an expression of personal or political differences. Nevertheless an interesting discussion occurred with the dissidents suggesting that Genetics Forum was failing to integrate disparate issues and networks into a coherent network or organization. Hence Les Levidow suggested a book on biotechnology as a whole to give a theoretical coherence to Genetics Forum as the only organization working right across biotechnology. Several speakers resisted his approach. Julie Hill argued that being a loose coalition was a strength, since an organization with a particular line is predictable and can be *'dealt with and absorbed by government'*. The coalition, because it is unpredictable, is able to *'hit and run on many issues. Lacking a grand strategy is a major strength, Genetics Forum needs an inchoate approach rather than a rational one'*.

The reason for co-ordinating a number of coalitions of different aspects of biotechnology was that in each coalition a different range of NGOs could be enrolled into an issue-specific network on an issue of mutual interest, without falling out over irresolvable differences they may have had on other issues. The intention was that the constituencies these NGOs speak for could be combined. Genetics Forum's anti-patenting coalition, Patent Concern, was recognized as a national umbrella group with 22 member organizations[7] opposed to patenting

---

[7] As of 24/4/95 the membership organisations of Patent Concern included: Pesticides Trust, Greenpeace UK, Green Group of MEPs, green Network, BUAV, UK Food Group, The Ecologist, National Food Alliance, Genetic Interest Group, Soil Association,

from a variety of points of view. For example, the Genetics Interest Group (GIG) worked on behalf of people suffering from genetic disorders and those who cared for them. They supported genetic manipulation, but opposed gene patenting because they saw it as slowing the development of genetic therapies to diagnose and 'cure' genetic disorders[8]. By contrast, the British Union Against Vivisection (BUAV) opposed biotechnology as a whole, as an animal welfare issue, with the patenting of Onco-Mouse as symbolic of anthropocentric exploitation and violence against animals. While GIG in a minority amongst NGOs in Patent Concern, they reflected the views of a liberal medical lobby which was crucial in the defeat of the Patenting Directive in 1995 (Interview, 27/6/95). These kinds of differences in the network made it impossible to articulate a single line on the patenting of human genetics, other than broad opposition. On plants, the split between biotechnology as unnatural and patenting as unjust, remained a problematic division (Meeting, 21/6/95).

Patent Concern did not hold a meeting in the time I was studying it. When a technicality in the EU procedures opened an opportunity to influence the political progress of the European Patenting Directive, Steve Emmott, the co-ordinator of Patent Concern, wrote a leaflet that was used by members of Patent Concern (and interested others) as the basis for lobbying MEPs. That is, after a long period of hibernation, Patent Concern was used more as a telephone tree to mobilise action than a meeting point for interested NGOs.

A coalition was set up to prevent the marketing of GM foods. The initial meeting (26/1/95), concentrated on three initial cases: tomato paste, oilseed rape and soya. The coalition included Greenpeace, the Green Alliance, Farmers' link, the Farm and Food Society and the Green Network as well as two representatives of consumer organizations. Each of the cases had its complexities leading to long debates over the possible environmental damage. Could outcrossing with weedy relatives occur in the UK, in Europe, or only in the USA where, they were to be grown? The low levels of immediate risk involved in these cases stood in contrast to the cumulative risks posed by their role as softeners for broader programmes of GMO release and marketing. Once again the NGOs were obliged to try and translate these open ended cumulative risks into technical arguments against specific cases.

WSPA, RSPCA, National Federation of Women's Institutes, SAFE Alliance, Compassion in World Farming, WWF (UK), Farm & Food Society, National Housewives Association.

8 The whole notion of 'cure' is of course deeply ambivalent, given that in the main diagnosis leads to abortion, and therefore to the elimination of disabled people rather than ending their physical suffering (Ruth McNally, Pers. Com. 1995).

The institutional fragmentation in the regulation of GMOs led to complex bureaucratic machinations, border wars and gaps in responsibilities and regulations. So NGOs were caught between Kafkaesque energy-sapping regulatory politics and public apathy ranging in the mid-1990s from passivity at best to indifference at worst. Thus the problem was how to escape bureaucratic politics, where NGOs were at a disadvantage. Mass participation in direct action at the time of the meeting seemed a remote possibility. Only when GM soya hit the supermarkets did vague unease erupt into widespread opposition. Robin Jenkins argued that consumers were the most influential constituency, hence NGOs could wield the nebulous sense of consumer unease as a lever on supermarket policy. The argument put forward at this meeting was that supermarkets were vulnerable to 'intelligent lobbying' from NGOs to guard against 'bad publicity', as has subsequently been demonstrated. Thus the central aim of the Genetics Forum workers was to agree a policy aimed at influencing the six major supermarket chains in the UK not to stock GMOs until the UK government enforced labelling and adopted clear regulations on environmental release of GMOs. Biomato paste or products containing GM soya would be the test cases. This strategy had already worked well with irradiated food, which had been approved for marketing by the state, but the supermarkets were convinced not to stock it.

Greenpeace: Global Reach

Greenpeace International has a global presence with approximately 4 million paying supporters world-wide. It has been called a 'transnational NGO' (Jamison and Eyerman, 1991) and likened to the Catholic Church for its global reach and mediated relationship of 'moral dependence' that extends well beyond the paying supporters (Szerszynski, 1997). Greenpeace has been characterized by a close relationship between the International executive and the 26 national offices, each with their own support base. Campaign objectives are set at the international level.

> *An annual general meeting where there are representatives of all the different offices get together to work out which issues Greenpeace will campaign on and how they will do it, which countries and so on and so forth* (Interview, 8/7/1994).

Greenpeace UK was set up in 1977 and by the mid-1990s had approximately 400,000 paying supporters (Frisch, 1994). Some of these supporters were organized in local groups, which fundraised and campaigned locally, but they did not determine the direction of the campaigns. Indeed, each Greenpeace national office was legally constituted as a limited company and even its paying supporters were not legally members of the organization, and therefore had no right to call the employees to accountability. It also held the intellectual property rights over the Greenpeace name as a registered trademark with a willingness to prosecute any local group that claimed to use the Greenpeace name without authorization from the national office (Rose, 1994). It was however unable to prevent the continued operation of the earlier Greenpeace (London), founded 1970 (Frisch, 1994). Greenpeace (London) was an anarchist collective active in a wide range of direct actions such as Stop the City in the 1980s, the anti-roads protests and the long running McLibel case in the 1990s. As a global organization, Greenpeace (International) had suffered from internal tensions between North and South, campaigners and bureaucrats, cuts in funding and restructuring with numerous job losses exacerbating tensions.

Greenpeace was heavily referenced in my snowball sample as a major group active on biotechnology, due to the presence in the UK of the then Science Director, Sue Mayer. Greenpeace was seen by some other NGOs as having a hard line against genetic engineering, patenting and GMO releases. This position has been likened to the 'German position' (the uncompromising stance of the German Greens), which it was argued by some, would drive biotechnology companies to relocate in the South where regulations were less stringent (Interview 6/4/1994).

Greenpeace focused on the environmental consequences of genetic engineering and consequently did not have policy positions on human or medical genetics. The two key issues for them were the release and patenting of GMOs. Only two national offices were working on biotechnology - Switzerland and the UK. Greenpeace opposed genetic engineering and the environmental release of GMOs on principle as an environmental risk (Interview, 8/7/1994). Patenting was a problem for them in that it would encourage genetic engineering and releases. The social justice dimension of patenting creates more ambivalence. The desire by Southern NGOs to develop compensation mechanisms for access to genes generates sympathy.

> *There's all that sense of injustice and things that you have in you and you think, actually, it's not right these people getting ripped off,* [But] *if the justice argument is just going to say well as long as we get enough money back, you know, you can plunder our genes and do what ever you want with them, that would be a problem for Greenpeace. You can't mess with genes anyway* (Interview, 8/7/94).

Similarly, Greenpeace campaigner Anna Brindley, called the fall of the patenting directive in 1995 'a moral victory for nature' (*New Scientist*, /3/1995) suggesting that biotechnology poses a technological risk to nature, not a problem of resource distribution. In other words there is a potential for a conflict of interests between a purely 'environmental' frame and social justice frame, although the two are seldom spelt out as stark alternatives in any position.

At the international level Greenpeace were in the front line of a campaign to have a Bio-Safety Protocol included in the Biodiversity Convention. Greenpeace were able to mobilise the G77 group of developing countries as well as the Nordic countries at the Biodiversity meeting in Nairobi in February 1994, to oppose a proposal from the UK and the Netherlands to drop the Biosafety Protocol in favour of voluntary guidelines. The Protocol has subsequently been adopted.

The political department of Greenpeace was responsible for lobbying. This involved pro-active work to link potential allies together around a position Greenpeace wishes them to adopt. However, it depended on the supply of scientific expertise by the science unit to make the arguments. Hence the science unit

> *does help to develop strategy and tactics using science and does interpretative work, produces reports and informs the rest of the office about what the science is and what it means in ordinary speak* (Interview, 1994).

Greenpeace material on the TRIPs was jointly written by Sue Mayer and Isabel Meister, in the Greenpeace International office in Amsterdam, and then passed on to a member of the economic department, Colin Hines, who would attend GATT and WTO meetings.

The science unit drew in expertise from sympathetic scientists working in universities or state research laboratories. The resulting scientific advisory network consisted of access to over a hundred scientists. Only a few were likely to be used at any one time depending on the nature of the political work at any particular time. These scientists may write or review reports or answer questions over the phone or come into the office for meetings or to brainstorm. The network is *'built up largely by personal contact, personal approaches, been to conferences, because they need to be sympathetic'* (Interview, 8/7/94). For example, Sue Mayer recruited John Porter to the science advisory network on the basis of personal friendship. Both used to work for Bristol University, Sue Mayer at the Langford Veterinary School and John Porter at the Long Ashton Agricultural Research Centre. Langford was working on respiratory diseases in cattle. When Bristol University accepted Ministry of Defence funding on the grounds they limit the work to an organism that was also a human pathogen Sue Mayer resigned her lectureship and accused the University of undertaking biological warfare research. John Porter was drawn into the campaign as he was on the executive committee of Scientists for

Global Responsibility (formerly Scientists Against Nuclear Arms). After moving to Greenpeace, via the RSPCA, Sue Mayer used John Porter's expertise to challenge patenting of genes by biotechnology companies in the European Patent Office on technical grounds (Interview 28/3/94). Plant Genetic Systems (PGS), based in Ghent, had applied for a patent on a genetically modified sugar beet, which it claimed was resistant to the herbicide, Basta, which was produced by the German agro-chemical company, Hoechst. The key technical argument revolved around clause 53B of the European Patent Convention that states that neither plant varieties nor essentially biological processes can be patented. Greenpeace argued that genetically engineering a change in a variety simply produces a new variety and that herbicide resistance is a biological process. PGS claimed that the use of a virus as a vector meant that the genetic change in the sugar beet was a micro-biological process and therefore patentable. The EPO found in favour of PGS but allowed Greenpeace to appeal (Interview 28/3/94). The EPO decision '*raises the intriguing possibility that if you introduce a gene into a whale or an elephant by ... using a virus, that it somehow become a micro-organism as a result*' (Interview, 28/3/94). Greenpeace subsequently won the appeal in February 1995, on the grounds that Basta resistant sugar beet was indeed a variety and therefore unpatentable. However, the president of the EPO Appeal Board referred the case for a further hearing, by the Enlarged Technical Committee of the EPO, to be held in private. In the meantime the EPO did not implement the decision against patenting varieties (Meeting, 28/9/1995).

As a civil servant, John Porter was allowed to give written evidence to the EPO, but prevented from appearing in Munich to give verbal evidence before the EPO hearing. Sue Mayer had been negotiating with Tom Blundell, the head of the Biotechnology and Biological Sciences Research Council, over the clear imbalance this represents in the access industry and NGOs have to advice and support from publicly funded scientists.

The very existence of the Greenpeace science advisory network and Scientists for Global Responsibility needs to be seen against a background of increasing commercialization and patent-driven research within the public sector and also international tensions over patenting. Craig Ventner in the USA was driving towards patenting as much of the Human Genome as possible while Charles Aufry in France was making his team's research publicly available on the internet and circulating a petition for scientists to oppose the patenting of their work (Interview, 28/3/94). Third World Network and Scientists for Global Responsibility had been circulating similar petitions.

Coalition politics, so central to Genetics Forum and the Green Alliance was marginal to Greenpeace, which preferred to work alone.

> *It's extremely unusual for Greenpeace to work in those kind of coalitions and one of the reasons is because we were always so uncompromising. To enter into a kind of compromising positioning with long drawn out discussions with other NGOs because usually, we find that isn't terribly effective way of using time and resources really. It's best to kind of go out and do something else with it to actually put your own message across. But ... particularly on new issues or where you do have a similar positioning, [it] can be extremely useful to join forces, only on an ad-hoc basis* (Interview, 8/7/94).

In the field of genetic engineering *'The only outside alliance stuff has been through Genetics Forum, Patent Concern'* (Interview, 8/7/94). This did not mean that Sue Mayer was not aware of other NGOs working in the field. She attended meetings of other NGOs, such as Gaia and Genetics Forum. It underplayed the link with the Green Alliance, which was used for a more conciliatory kind of work than usually characterises the Greenpeace style.

GRAIN / RAFI: From Founding Father to Movement Leaders

GRAIN (Genetic Resources International) was referenced by everyone else in my sample as the acknowledged experts in the field of seed politics[9]. Two organizations lead the global anti-GM movement in research and campaigning: RAFI in North America and GRAIN in Europe. Both these organizations owe their origins to a single individual, Pat Mooney, who was recognized as the 'Founding Father' of the anti-GM movement. Mooney displayed the classic characteristics of a charismatic leader.

> *He's a very brilliant individual who gets his kicks out of doing very detailed research and expressing it in a rather interesting way and that research is useful. He also has a strong sense of mission and some people think he's quite a sort of charismatic figure in this world, in this sort of milieu* (Interview, 3/3/1995).

In 1977 Sri Lankan tea plantation workers alerted Pat Mooney to the loss of varieties in the plantation system (Pers. Com., 14/6/96). In response, he convinced a group of European NGOs, the International Coalition for Development (ICDA), to form the ICDA Seeds Campaign in 1979. Pat Mooney left after a dispute with the ICDA board in 1983 and set up his own NGO in North America, RAFI (the Rural Advancement Foundation

---

9 Although it is based in Barcelona, I was able to interview Dorothy Myers, a former chairperson of GRAIN, living in Oxford, on 3/3/1995. She is the former chairperson of GRAIN as well as the Pesticides Trust and has worked for Oxfam. She was then working part-time for the Pesticides Trust on the Organic Cotton Project.

International), with its head office in Canada. Meanwhile Henck Hobbelink took over as the ICDA Seeds campaigner. However, tensions between ICDA and the Seeds Campaign remained. Hobbelink was able to raise money on the basis of the Seeds work from other big NGOs, but ICDA controlled the money and used some of it for other purposes. Eventually in 1990, Hobbelink took the Seeds campaign out of ICDA to become a separate organization called GRAIN, and moved the whole operation from Brussels to Barcelona. The style of work of the two organizations has been contrasted as follows

> *Pat Mooney, who calls himself the director of RAFI, but he isn't the director, really, of anything. An organization doesn't really exist. They sort of have a board. GRAIN isn't like that. GRAIN is much more of an organization. It functions as an organization, as a collective with a defined purpose and the usual kinds of support mechanisms, with a staff and a board, finances and practices* (Interview, 3/3/1995).

In short, RAFI still displays the charismatic authority of its founder, while GRAIN is more rationalized. Their research styles have also been contrasted, GRAIN with a more European historical orientation, RAFI more North American, strategic and accessible to journalists (Pers. Com., 18/6/96). Much of the information and analysis other NGOs use is derived from either GRAIN's *Seedling* or RAFI's *Communiqué*. Pat Mooney is still the source of many new terms such as 'biopiracy'. In spite of some tensions between Pat Mooney and GRAIN, a GRAIN / RAFI axis is clearly visible at the centre of the anti-GM movement.

GRAIN continued to expand through the 1990s, drawing in board members from several European countries including Miges Baumann (Swissaid and now GRAIN) and Michel Pimbert (WWF International, then WWF Switzerland) as well as Dorothy Myers (Oxfam UK). Unlike RAFI, the GRAIN board members had all been active players in genetic resources through their own NGOs, and worked closely as a team. So that the experience *'wasn't really like a staff-board relationship, just a group of individuals working on the issue'* (Interview, 3/3/1995).

Funding for GRAIN came initially from development NGOs, but gradually shifted to governmental aid budgets. GRAIN emerged from a development nexus and retained its North-South focus. GRAIN's role in relation to genetic diversity loss in European agriculture, and local seed saving projects, has been a vexed one. Although a GRAIN employee wrote a book on this topic, (Vellve, 1992), it has never been central to their work, a point on which the NGO community as a whole has been criticized by a practical seed saver (Interview, 21/3/94). GRAIN see farmers as their constituency and seeds as their issue, whereas RAFI have widened their brief to take up indigenous people and the questions of IPRs over pharmaceuticals and human genes.

Dorothy Myers identified a number of strands to the work GRAIN does, all of which are global, '*there are all these strands to the work, there's the Convention on Biodiversity, there's the whole GATT - TRIPs stuff, there's the reform of the CG system*' (Interview, 3/3/1995). The Consultative Group for International Agricultural Research, known as the CGIAR or CG system was set up in 1971 to support the Green Revolution. CGIAR included 18 International Agricultural Research Centres (IARCs), with a combined *ex-situ* collection of 4,500,000,000 seeds. They held the definitive collections of all major food crops such as rice (Philippines), maize (Mexico) and wheat, and other commercially important crops such as cocoa (Trinidad) (Baumann et al, 1996). All these collections were excluded from the provisions of the Biodiversity Convention. The CGIAR system, which had been run as a private club by the major (English speaking) donors: the USA, Canada, UK and Australia, became the subject of a battle for control between the FAO and the World Bank in 1994. The FAO, somewhat surprisingly, won control (GRAIN, 1994; *New Scientist*, 2/7/94 and 5/11/94).

> *The CGIAR system is ... in a state of crisis at the moment. A lot of people would like to see it reformed, in order to take better account of farmers' real needs, to take a much more bottom up approach to its research. The IARCs ... drove the Green Revolution and there is still a lot of Green Revolution type thinking around in the IARCs. The CG [is] driven by the World Bank, [and] the big donors, but there are a lot of people around who would like to see that change. I mean the Robert Chambers types, GRAIN, RAFI, a lot of Third World organizations* (Interview, 3/3/1995).

> *FAO, at this point is more or less given over to the NGOs' agenda. A lot of them have joined, on the staff, and in other ways in FAO, but the point is, FAO is no longer the big heavy powerful organization that it was, its becoming a sort of the technical arm ... of the World Bank, more or less. Now the World Bank is the big operator now, and FAO is now sort of NGO dump ... The dynamics now are much more to do with getting the World Bank to change, and the Americans and bilateral [agreements]. That's where the real battles are to be fought...And all that's happened in the last fifteen years. So right through the 80s the relations between FAO and NGOs was completely revolutionized. I mean NGOs were sort of shut out of FAO in the early 80s but gradually they were becoming increasingly accepted and had done serious work and made serious contributions. That's been accepted, taken on board, valued and now we're at the point that there are NGOs or former NGOs actually on the staff of FAO. Cary Fowler, who was outside pissing in, is now inside pissing out. He's now actually in charge of the arrangements for the technical conference in 1996. I mean there are other NGOs that are there as well. So it's a whole new thing at*

> *FAO. You know there aren't still many battles there to be fought* (Interview, 3/3/1995).

NGOs, such as GRAIN and RAFI, have become effective actors within global institutions, whether or not they can count on the support of their own governments.

> *I've done it myself ... you interact with the delegates and with staff of FAO and your own delegation is kind of irrelevant. Because they will be taking a position you don't support. You go there and you work with or through colleagues. You work with delegations, which you think are going to take your line. I mean the most effective time that happened was when we were trying to get the prior informed consent through the FAO. Well we got the code adopted. By working with partners from Latin America, Asia, Africa and we got NGOs from those countries to come and work with their own people, their own delegates, and the delegates from the region. Eventually we got the code adopted. Then we produced another piece of research and passed that out. Again got people from the regions. There was a big co-ordination to get people from the regions to come and lobby their governments. We people from the UK weren't doing a lot of lobbying our delegate. It was a waste of time. So we put our efforts into facilitating and co-ordinating efforts by others. Going, as you say, over the heads, through our own national delegate... it's a feature of the way things happen now* (Interview, 3/3/1995).

NGOs had proven themselves as counter-experts, operating successfully within the system.

> *You need a high level of expertise to engage with the official system. How do you do it on these kinds of issues except by acquiring a high level of expertise? As much expertise as the official, as much or more expertise as the officials, who are having to deal with it, set policy* (Interview, 3/3/1995).

The network of NGO actors in the UK, directly concerned with the control of genetic resources, in Dorothy Myers opinion, was growing but remained small:

> *There is quite a collection of people with an interest. I mean more so than even 5 years ago. There are probably 10 or a dozen people now who are interested. So all these people, they're individuals really with an interest, and in some cases they are bringing the organization in behind them. It's to do with key individuals who have the knowledge and interest and then try to persuade their organizations to do something* (Interview, 3/3/1995).

The network was a hybrid, simultaneously a public network of NGOs and as a personal network of individuals. While most of the individuals involved were based in NGOs, the network could not be reduced to interactions between organizations with established positions and interests. The network consisted of small organizations, often indistinguishable from the individuals who run them, on the one hand and on the other individuals within larger organizations that have a greater or lesser influence over the rest of their organization. One set of connections, which could have been closer, was that between the campaigners and those directly involved with seed saving.

> *They [are] all aware of each other, [but] you certainly don't see the seed saving activities translating into anything much more political I don't think, but they are a bit interested in what's going on internationally, but not that much really. Like you get this woman, Nancy Arrowsmith, who does this kind of activity in Austria, who grows things. She's quite active politically at an international level. It's part of the whole European thing, which could be stronger* (Interview, 3/3/1995).

Farmers' Link: Bringing It All Back Home

Farmers' Link was unique among the NGOs examined in this chapter, in that, it combined a local base in the Norfolk farming community with addressing global issues[10]. It started out as a spin-off from the Norwich Third World Centre, but outgrew its parent organization and became a separate NGO with an office in Norwich. Its initial concerns were launched in 1986 at a joint meeting with the Norfolk section of the National Farmers Union entitled *Norfolk and Third World Farming: What are the Links?*

The locality contained not only farmers, but also,

> *we were sitting on 850 plant scientists, the biggest concentration of them in Europe, along with Cologne, and we couldn't ignore the fact that these people were some of the pioneers of biotech'* (Interview, 6/4/1995).

The new Vice Chancellor of the University of East Anglia, Derek Burke, had been recruited from a Canadian biotechnology company. His project was to link the University much more closely into the work of the related research institutes in the area. The John Innes Institute was particularly eminent, especially after it absorbed the privatized Cambridge Plant Breeding Institute.

Mindful of the biotechnological resources on their doorstep, Farmers Link sought to negotiate the terms on which biotechnology would impact on agriculture.

---

[10] This section is based on an interview with Alistair Smith on 6/4/1995.

> *To actually raise an informed public debate about a subject which was either Frankenstein's monster or the best thing since sliced bread, with not much in between. From 89/90/91 we were quite heavily involved in the debate and were for that period the major NGO actor lying somewhere between the out and out anti-genetic engineering, anti-biotech type stance and those who were completely uncritical. We were in a space in the middle which nobody else was occupying, certainly not in Britain* (Interview, 6/4/1994).

They hoped to avoid what has been called 'the millenarian' alternatives (Haraway, 1997) of naive techno-optimism or critical despair. Instead, they wished to develop a *'more refined critique'* than was available, distinguishing between the possibilities of appropriate agricultural biotechnology and the dominance of vested interests. While appropriate biotechnology research was continuing (for example by Guido Ruivenkamp in the Netherlands and Swaminathan in Madras, India), it soon became clear to Farmers' Link that, as they had suspected, the biotechnology industry was driving agricultural biotechnology towards purely commercial ends. Alistair Smith observed the commercial forces at work in the international diplomacy of biotechnology and biodiversity when the CGIAR seed banks were removed from the biodiversity negotiations during the Rio process:

> *deals were done two weeks before Rio, in Istanbul, whereby basically all the genetic material that counted in terms of future strategic food resources were taken out of the remit of the Biodiversity Convention* (Interview, 6/4/1994).

From 1991, the work of Farmers' Link moved beyond North-South links. Extensive contact with Southern NGOs redirected them to 'shift the Northern landscape'[11] that is, work towards sustainability in the North. Hence Farmers' Link recruited farmers to a working group to explore the meaning of sustainable agriculture in Britain, including participatory farmer research. This elusive research paradigm, which was yet to be developed in any detail, had to be as dependent on social and cultural determinants as much as genetic factors. It was

> *bringing in all the conventional, new conventional development wisdoms and participatory farmer research and all this stuff that people spout about and very few people do - bringing it to the North'* (Interview, 6/4/1994).

An obstacle to sustainable agriculture in the North was the link between genetically engineering herbicide resistance and marketing herbicides. Once

---

[11] This term was coined by Filipino activists (Interview, 6/4/1994).

again on their doorstep in Norwich is Rhône-Poulenc the French agro-chemical transnational, which produces the herbicide promoxinol in Norwich. 'All the promoxinol in the world is produced in Norwich' (Interview, 6/4/1994). Rhône-Poulenc has a deal with the American biotechnology company, Agracetus, which has developed a cotton variety resistant to promoxinol. Alistair Smith was able to interest Greenpeace in the promoxinol affair from a toxic trade point of view[12].

However, he distanced himself and Farmers' Link politically from any in principle blanket opposition to biotechnology articulated as either a Southern position by Vandana Shiva, *The Ecologist* and the Gaia Foundation, or from a more conventional Northern environmental risk perspective by Greenpeace. Despite expressing great respect for Sue Mayer, he described the Greenpeace position as ambiguous.

> *It's a bit of a moot point in the NGO community as to what Greenpeace's policy actually is. Nobody really knows, possibly neither do Greenpeace. Their great temptation is to go down the German road of getting bans on this, that, and the other, which they must have spotted, without parallel action in the South, essentially just forces the companies to migrate* (Interview, 6/4/1994).

Instead, he advocated attempts to produce an appropriate plant biotechnology through a close understanding of rural poverty, and hence a concern for the effects of biotechnology on the distribution of ecological resources. However, the justice argument is closely connected to maintenance of biodiversity. Biotechnology companies claim biotechnology is a means of conserving biodiversity.

> *These are tools with which we can increase (a) our methods of conservation and (b) the gene pool. I think it's a load of crap myself. That seems to me to be the key question. Very few people are working on it and it's a key claim of the proponents of Biotech* (Interview, 6/4/1994).

Along with appropriate biotechnology, appropriate IPRs were needed to protect peasant farmers. In India attempts were being made to produce forms of IPRs, which are both appropriate to peasant farmers and fit into the WTO as *sui generis* systems allowed as an alternative to patenting.

Within UK politics, Alistair Smith had resisted participating in dialogue with government, organized by Julie Hill. While he approved of the idea of

---

12 Alistair Smith wrote this up for the Greenpeace Toxic Trade Journal, with then Greenpeace campaigner, Topsy Jewel (Interview, 6/4/1994).

making Julie Hill's position on ACRE more democratically accountable, he considers it comes up against

> *a massive information gap. Participatory research on Biotech stuff is really a non starter at this stage of the game because there are so few farmers, or any of us, really know enough of the science. The fundamental problem that we have in this country is the problem of the nature of the science, the complexity of it and the state of supposed democracy* (Interview, 6/4/1994).

Farmers' Link appointed Paul Farbon to take up their biodiversity work and he appeared in all the meetings and seminars for the rest of my study period, except Leipzig. Paul Farbon was seconded to Christian Aid to work on One World Week, which attempted to widen the scope of the seed debates and make biodiversity into a popular grass roots issue in the UK.

Intermediate Technology: Southern Seed Saving and Northern Counter-Expertise

The Intermediate Technology Development Group (IT) was set up in 1965 by E.F. Schumaker, author of the Green classic *Small is Beautiful* (Schumaker, 1974). Its purpose was to promote intermediate forms of technology in the South, which fell between the indigenous low tech associated with poverty and hard labour and imported high tech associated with technical and financial dependency and the creation of comprador technological enclaves. The argument was never a linear one about finding a mean between extremes. Rather technology was to be framed within a set of social and ecological imperatives. The social situatedness of technology was to be expressed in terms of its appropriateness to the context. IT itself still retained its head office in the UK, but it also had offices in Peru, Zimbabwe, Kenya, Sri Lanka and Bangladesh and had shifted much of the decision making to these national offices.

Within biodiversity, IT maintained an emphasis on the way in which biodiversity was utilised in subsistence by farmers, pastoralists and fishers. Biodiversity included domestic animals and fish species, as well as crop diversity. For Patrick Mulvany[13], biodiversity had to be *'dynamic diversity'*. Firstly, diversity needed to be conserved *in-situ*, in the fields, not abstracted into gene banks (or *'gene morgues'*), where seeds lose their adaptation to field conditions.

---

[13] This section is based on an interview I did with Patrick Mulvany on 9/5/95.

> *The NGOs and lots of them are now quite active in looking at strategies which will promote this in situ conservation and development of diversity. At root it is much more an issue about diverse agricultural policies than anything very scientific about genetic reserves as such* (Interview, 9/5/1995).

Secondly, dynamic diversity could not be achieved by a static approach that forced peasants to adopt a rigidly traditional set of practices, for example as the condition of receiving protection in an extractive reserve. Dynamic diversity signified biodiversity as a part of creative cultural innovation, combining past resources and practices in new ways to respond to changing conditions and tastes.

For example, IT promoted seed fairs in the arid Masvinga province of Zimbabwe. These 'communal areas' were set up in the colonial period deliberately as areas of food insecurity; to supply labour to white owned farms, mines and industries. Maize had largely driven out small grains such as sorghum and millet, even though they were indigenous to Africa and far more drought resistant. Maize was higher yielding with sufficient rain, softer and easier to mill. It was also the primary form of relief food available. Maize came to stand for modernity in rural Zimbabwe. The Zimbabwean Government, concerned with the chronic state of food insecurity, consulted IT.

> *So, all of those things conspired to make maize much more popular as a seed, but what was interesting was when these particular communities had gone through a process of analysis of what they required and the kind of cultivation techniques and water conservation techniques and so on, the next set of questions were about germplasm, about seeds. People starting remembering... "Dad used to sow this, Grandpa used sow that and Grandma has these squashes and things in her garden" and they start asking questions about it. For the last three years my colleagues have helped the community organize seed fairs at which they brought all the varieties that they value to be exchanged amongst themselves. Something of the order of 50 varieties each time have been produced from the communities. They've now got lists of things they would like to see, they could remember which probably aren't found in the locality[any more] and ways have been found to look out these varieties. So, a tremendous sort of resource has suddenly appeared out of back gardens and little corners of fields. More than that it was so exciting to walk around the fields in company of various farmers, for them to point out what they were experimenting with and how, the mixtures of varieties and places they plant. What was being released was this creativity for diversity in the production system. I guess that's influenced me probably more than anything else in the last few years. If that kind of thing can be released in this tiny little microcosm of an experiment, if that kind of approach was supported as agricultural policy you're on a winner. That's how food security happens, certainly at the household and community level* (Interview, 9/5/95).

IT's work on this kind of sustainable agriculture sharpened its political profile in the run up to Rio and the Biodiversity Convention. IT organized the *Gene Traders Conference* in London in early 1992, which proved to be an important marker in the consolidation of UK networks around agricultural biodiversity. After two years in a *'fog of re-organization'* Patrick Mulvany returned to the issue in 1994. He saw the period up to the end of 1997 as a key period in which international regulation would be *'set in a particular direction which will be more a corporate line or more of a civil society line'* (Interview, 9/5/95).

An important event in this re-organization of the international agenda was, in his estimation, the FAO's Fourth Technical Conference in Leipzig in June 1996. Patrick Mulvany set in motion a new network under the name 'Food for Life', drawing together the most closely focused group of NGOs concerned with seeds. It was re-named the UK Food Group, and after Leipzig became ABC (Agro-Biodiversity Coalition). ABC became a specific grouping inside the wider UK Food Group at the FAO's World Food Summit in Rome.

Gaia: The Spirit of Networking

A member of the Gaia Foundation defined it as, *'A European base for a number of individuals and networks in South America, South East Asia and Africa committed to maintaining cultural and biological diversity'* (Meeting 21/6/95). They have also, somewhat ironically, been described as *'Deep Greens in search of an issue, who think biotechnology may be their issue'* (Interview, 5/5/95). While Gaia's rationale was to support Southern groups, especially indigenous people, their contact with these groups also gave Gaia a Southern power base from which to exercise some leverage on UK and other Northern NGOs.

The Gaia Foundation operated by bringing together eminent people. Gaia events had a particular quality of conviviality. For example in July 1995, Vandana Shiva was invited to address a small meeting organized by the Gaia Foundation, with about 70 invited guests in attendance. It was not an open public meeting, nor a closed business meeting, but more a party for a select gathering of environmentalists.

The gathering started as a buffet supper, with lots of chatting and networking, at Gaia House, a terraced residence in Hampstead, with the feel of a comfortably elegant domestic home. The lower ground floor room opened onto a back patio and lush garden, providing an appropriate setting for relaxed socialising. The meeting proper followed in a local hall across the road. The meeting room again evoked a relaxed Hampstead elegance with wood panelled walls, decorated with ferns and 'Best London Garden' awards. The choice of these settings was significant. Every discourse or network is context bound.

The ambience created through the physical nature of Gaia House and its immediate environs, gave a particular grounding to Gaia's discourse. Attention to the aesthetics and partnership with nature through gardening, implied a Green spirituality in tune with nature not only in its threatened fragility, but also in its bountifulness, made accessible by social privilege. In contrast to more instrumental meetings, where NGOs were required to negotiate around agendas and formulate policies, the Gaia meetings were more of an opportunity to recharge and re-establish a community of feeling. Gaia was able to draw in representatives of more instrumentalist NGOs, such as Greenpeace, WWF and Farmers' Link, as well as the late James Goldsmith MEP, and the New Economics Foundation.

Vandana Shiva, in addition to an NGO of her own, was a prominent member of the Third World Network and a heavily published author. She was the best known Southern intellectual working on seeds, biotechnology and IPRs. She spoke on globalization and local resistance in India.

Her key point was that local communities were the principal sites and agents of struggle against the globalization of the economy. Local communities took direct action to prevent the arrival or departure of TNCs (including Cargill Seeds) from their locality. Direct action was opposed by the Indian military and frequently led to loss of human life. Thus with the loss of any national protection local communities were exposed to unmediated conflict with global power. However, in response to questioning, she revealed that local resistance derived crucial support in the form of information (or intelligence) from Northern NGOs, with faxes flying back and forth between activists in North and South.

Shiva's rhetorical project was to set up a central political dialectic between globalising Northern TNCs intent on enclosing the commons, and local communities in the South protecting the local environmental commons. This presentation erased the mediating role of Southern and Northern NGOs in framing the terms of conflict, and so also unreflexively erased her own contribution to the framing process. Support from Northern NGOs, (and her own transnational status as the North's favourite Southern intellectual), undermined the simplicity of the opposition of the global versus the local. Clearly antagonisms at each end of the global-local divide are reproduced and transmitted through the particular conflicts. No locality is occupied by a single unified community. Local communities are sites of multiple identities and complex power relations - class/caste, ethnicity/religion and gender being only the most obvious cleavages.

A more select Gaia meeting on food politics was held with Vandana Shiva and chosen UK NGOs the day after the gathering. It is in these smaller, less accessible meetings that decisions are made. Both Shiva's charismatic presence and the Gaia's style of informal networking were used to great effect in the NGO mobilization in Leipzig in June 1996. There Liz Hosken, Gaia director, effectively acted as broker, bringing together the RAFI / GRAIN axis with the Third World Network and Latin American NGOs, into a global alliance, led by Pat Mooney, Henck Hobbelink, Vandana Shiva and Liz Hosken. Gaia cuts the deals for Shiva to speak on behalf of a global anti-GM movement.

WWF: Unwilling Host

WWF was the second global NGO widely cited in my sample. It was principally WWF (International) in Switzerland, that was seen as a major player in biodiversity, rather than WWF (UK). Much, however, rested on the reputation of Michel Pimbert, who organized an international symposium on biodiversity and patenting in Bern. (See next chapter). Following the symposium, Pimbert was sacked by WWF (International)[14]. The links that he forged between nature conservation and social issues within WWF (International) were to be severed. Other supporters of community-led, *in-situ* biodiversity were axed as well (Interview, 6/3/1995). Richard Tapper of WWF (UK) suffered a similar fate. He too, was made redundant in 1995, as a result of taking the organization further on issues of biotechnology and biodiversity than influential people wished to see it go (Pers. Comm., 11/6/96). As a global organization, WWF had played host to a separate seed network, which became expendable[15]. Henner Ehringhaus, former deputy director of WWF (International), described two schools of thought in the conflict within WWF.

> The old school is interested in animals, plants and protected areas. They are interested in people and development as an instrument to protect the forests. They are really very anti-people. The new school is interested in sustainable development, meeting human needs, helping people improve their agriculture, and working with local communities in a more integrated approach to nature conservation. What we have seen is those of the new school fired, and those of the old school promoted (Guardian, 18/11/95).

---

14 After a period of consultancy he has been re-employed as Director of WWF Switzerland.

15 Reports by Dorothy Myers of GRAIN and Farhana Yamin of FIELD have, however, subsequently been published by WWF.

HDRA Heritage Seed Library: Backyard Biodiversity

In the 1970s, HDRA set up a Heritage Seed Library near Coventry to collect varieties of fruit and vegetable that were rapidly being lost. Losses occurred partly through changing practices, but mainly in the wake of the 1964 Plant Varieties Act. This Act set up the national seed register, which made the sale of small scale unregistered varieties illegal. Further impetus came from the 1981 EU Seed List, which wiped another 1500 varieties off the market. The Seed Library itself was in a poor state until 1992 when the charismatic environmental journalist Jeremy Cherfas was appointed as its director. The membership of the Seed Library has grown continuously, reaching 4,500 in 1994 and over 9,000 in 1999. A seed exchange scheme has been attached to the catalogue and various local schemes have appeared, for example one run by West Somerset Organic Group. Jeremy Cherfas described his motivation as coming from the realization that vegetable varieties represented a 'biodiversity extinction crisis' that he could actually do something about (Cherfas et al, 1996). Cherfas's story of his own engagement with seed saving, amplified by presenting similar biographical accounts of other eminent seed savers, was woven into a movement story-line, within which his readers could locate themselves as seed savers.

Seed saving within HDRA followed a triple track system to fulfill three separate aims. The Seed Library acted firstly as an *ex situ* gene bank, preserving a base collection of genetic diversity in fruit and vegetables. By 1999, there were over 800 accessions in the base collection, which it grows out periodically on site at Ryton Gardens, in open-air plots and poly-tunnels. Plus 150 'pre-accessions', not yet available to members. Secondly it was a point of public access to plant genetic resources in an active collection. The active collection circulated between the 55 seed guardians and the wider membership. Each seed guardian took responsibility for a particular variety. They received seed from HDRA, grew it out and returned the multiplied up seed to HDRA, for distribution to the members. This active collection was kept separate from the base collection held and grown out at Ryton. Thirdly it aimed to encourage *in situ* conservation by gardeners. The ordinary members of the Seed Library ordered their 6 packets of seed from the 200 varieties listed annually in the catalogue. They grew out the seeds they received and saved them if they could, passed them on to others or used them to experiment with to produce their own hybrids if they wished.

With *in situ* conservation genetic drift may occur so that intentionally or unintentionally the seeds saved after several generations may have quite different genetic make up from those in the *ex situ* collection. This would occur simply through the selection of the most successful plants in a particular area, let alone through cross-pollination with relatives. *In situ* conservation is not the degeneration of the pure genetic stock held in the base collection. It is technological innovation, active biodiversity, of the sort which seed activists applaud in Southern peasant farmers. The Seed Library has identified 60 new distinct varieties through their 'Seed Search' campaign to acquire the heirloom varieties held in members' families. These were sometimes donated to the Seed Library precisely because traditional, local knowledge associated with their cultivation had died with an elderly relative.

Like the other NGOs, the Seed Library provided counter-expertise, lobbying on its particular area of interest, namely the loss of diversity in domestic vegetable and fruit varieties. Uniquely in this area it also performed a 'grass-roots engineering' role developing and promoting an alternative biotechnology. It is unexpected to describe gardeners as technologists. Yet technology always includes knowledge and practices as well as machines, or in this case genes. This backyard biotechnology consisted of a combination of genes, situated knowledge and practices as an alternative to the dominant commercial biotechnology of the transnational agro-chemical companies. The Seed Library was also unusual in having a large and often highly skilled membership engaged directly in its work. Its sharpest political contribution was to free forms of biotechnology from industrial and bureaucratic control and to connect it back to a wider popular base.

Counter-Experts, Leadership and the Public Interest

During the mid-1990s the anti-GM movement was still in an early phase. A core set of counter-expert activists was consolidating into the first generation of movement leaders. The wider popular mobilization against GM food was yet to emerge. A central problem for this leadership was how to keep in touch with its followers. As counter experts they used their limited resources to hammer out their arguments into an injustice frame defining the issues that concerned them. They had to identify their adversaries and engage with them, while claiming to represent a constituency. All too often it was extremely difficult to establish the identity of this constituency, let alone how to mobilise them. NGOs arise from civil society, separate from state and industry, and the experts aligned to them. NGOs were inclined to present themselves as representing civil society as a whole in the form of 'the public interest', rather than a particular constituency such as environmentalists. Indeed NGOs are sometimes referred to as 'public interest groups'. While it was relatively straight forward to show that the NGOs did not represent a 'private interest' in profit making, it was more complicated to show that they articulated 'the public interest', or the interests of defined publics. The idea that NGOs organically express the concerns of civil society can be tricky to sustain, especially as counter-experts have to learn expert languages, from which the public is excluded.

The NGOs were well aware of their position and reflected on it collectively. During the summer of 1995, the Green Alliance hosted a series of three seminars to feed social scientific perspectives on the social and cultural dimensions of risk back into the narrower policy debates. One of these seminars was a collective reflexive attempt to reconsider the wisdom of having become too deeply rooted in a counter-expert identity where it became easier to converse with their adversaries than their supporters. The discursive context in which the NGOs operated was described as a dominant paradigm of probabilistic technical risk assessment. This expert discourse claimed to quantify any risk and thus make it possible to measure it against the yardstick of the health risk of smoking. The narrow definition of risk within expert discourse excluded qualitative factors such as the cumulative effects of repeated risks, the irreversibility of certain risks, or the separation between those who produced the risks and those who were affected. The lack of fit between the narrow expert version of risk and the wider public version has taken the displaced form of oversized public inquiries, battles over safety regulations and endless debate over risks which the experts felt they had long ago demonstrated were safe (Presentation, 16/6/95; Grove White, *et al*, 1997).

The NGOs drew their support from the trust placed in them by the public. Such trust is based on the public's moral and cultural identification with the NGOs. When they argued with industry and regulators over technical risk assessment they did so as a vector of the more inchoate public unease. Yet, in order to contest the accepted risk assessments the NGOs found themselves translating cultural unease about high risk technologies into the technical terms demanded by the regulatory regimes. In this process of becoming expert, they were apt to lose touch with the feelings of their public.

At a subsequent meeting, organized by the Genetics Forum, NGOs struggled over whether their role was to reflect 'the public interest' or construct it. A Greenpeace representative favoured moulding 'the public interest', while a consumer representative was shocked at the fickle attitude of the public towards biotechnology *'most of them are more interested in East Enders'*. Julie Hill, of the Green Alliance, favoured reflecting the public interest, not moulding it, but suggested that *'they expect you to look after it for them'*. The implication of her observations suggest that the public placed its trust in NGOs and expected them to argue on behalf of the public on issues the public did not understand in any detail. Therefore NGOs could not simply reflect public interests, since these interests are not clearly articulated. They had to interpret public feeling, to extract concerns, rather than attempting to convince the public or ignore it. Public trust in the NGOs required them to take an advocacy role, but still to give a disinterested view rather than manipulating the public for organizational aims.

Nevertheless this counter-expert leadership seemed isolated from a wider public following in the mid-1990s. One possible explanation put forward for why the seed networks were still so small was that the complexity of the issue made it inaccessible to a wider movement.

> *This issue isn't well understood. It's an extremely complicated issue and that may be why it's interested individuals who have provided the energy, taken the lead, people who have had the time to get inside the issue. There isn't anything immediate, like veal calves being exported, that you can mobilise around* (Interview, 3/3/1995).

However, the issues were neither simple nor complex in themselves. Rather issues, which appeared as complex within arenas of policy formation, may be framed as simpler moral issues in wider arenas of popular mobilization. NGOs' lack of success in popularising seed biodiversity threats at a popular level contrasted strongly with their success in mastering a counter-expert role within the world of seed diplomacy. They participated in and constructed alternatives within the discourse of international negotiations, and even pushed back the boundaries of this discourse. What had so far evaded them was to connect their arguments to popular perceptions. Indeed, much of the anxiety expressed by NGOs concerned the greater ease with which they inhabited the worlds of expert regulation and diplomacy as opposed to the world of popular dissent.

The reasons for a social movement succeeding or failing to mobilise large numbers are the subject of extended debate, but one aspect is clearly the articulation of a 'diagnostic frame' (Snow and Benford, 1988) or 'injustice frame' (Gamson, 1995). This is a discourse that symbolises a social or environmental problem in a way that is accessible to various social groups, to make possible a set of networks and alliances. Such a frame needs to bring into play key symbolic terms that can be interpreted in a variety of ways by different social groups or networks within a movement. These symbolic terms must allow not only the combination of different constituencies, but also the translation back and forth between the counter-expert language of negotiation, and the multiple popular languages of social movement identities.

Charismatic mega-fauna, such as whales (Stoett, 1993), or mega-flora, such as oaks in British anti-roads protests (Welsh and McLeish, 1996) clearly have been able to play a significant symbolic role, which may be related to their great size. In spite of their small size, seeds too have their own metaphoric resonances, including 'seed' for sperm, therefore metonymically linked to male potency and 'to seed' for generation and creativity, expanded in the titles of key texts such as *Seeds of the Earth* (Mooney, 1979) and *First the Seed* (Kloppenburg,1988a).

Successful framing requires fidelity to wider culturally significant narratives (Snow and Benford, 1988). Yet key narratives are culturally specific. In France, the protection of small farming systems has provided the principal objection to patenting seeds. In Germany, the Nazi period has left ethical fears of the role of science in eugenics which has given the opposition to biotechnology a sharper edge (Interview, 20/10/1994). In Britain, it was the BSE crisis that generated intense distrust of experts from the food industry and government regulators that has transferred to GM food. Whether Third World peasants and their plants can come to occupy an important place in British hearts has remained an open question.

New Actors Join the Fray

Visible mobilization phases of social movements are frequently preceded by 'trigger events', which draw in a new type of movement leader, the 'rebels' (Moyer, 1987; 1990). Two types of 'trigger events' sparked the move from the latent phase of the anti-GM movement into a mobilization phase. First was the marketing of GM food in the UK. Second was the exhaustion of the mobilization phase of the anti-roads movement, which made available a contingent of rebels ready to shift to a new movement. Third was the network connections created at the Hunger Gathering organized outside the FAO Food Summit in Rome in November 1996, which drew in British direct action 'rebels' in significant numbers, who brought the issue home to the UK.

Since 1997 the number of NGOs interested in GM food, biotechnology, and patenting has grown, but the ways of framing the issues has not changed fundamentally. The mass membership environmental organizations, Friends of the Earth, a new comer to the issue and Greenpeace, previously restricted to a counter-expert role, launched campaigns. As did the Women's Environmental Network. The Genetic Engineering Network was set up by Reclaim the Streets, Earth First!, the Women's Environmental Network and Greenpeace (*GenEthics News*, 1997). The Soil Association too joined the fray, declaring that genetic modification was an obstacle to organic registration. Local GenetiX groups have appeared all over the country and repeated the debates in front of a wider more grass roots audience. The Soil Association became now a regular contributor in this circuit of anti-GM speaker meetings around the country, along with established anti-GM campaigners such as GenEthics News and more recent arrivals like the Women's Environmental Network and the Green Party (Bristol Genetix meeting, 29/4/1999).

By this time a wider mobilization against GM food, embodied in such organizations as the Genetix Snowball, connecting Earth First activists and the aficionados of the Glastonbury Greenfield, was emerging as Britain's newest Green protest movement. These networks were more inclined to taking direct action than the NGOs who provided the first generation of leadership for the movement and drew on the more recent waves of direct action against road building.

The first direct action against GM crops in the UK took place under the auspices of the GenetiX Snowball in July 1998 when 5 women uprooted GM crops from a Monsanto demonstration site in Oxfordshire. Genetix Snowball was set up as a Non Violent Direct Action campaign based on a similar snowball against nuclear weapons in the 1980s. While the first action led to arrests but no charges, subsequent actions followed and provoked injunctions and High Court Hearings (Genetix Snowball website, 2000).

These developments signalled an expansion in the networks and shifts in the action repertoire of the anti-GM movement. As new activists were drawn in from previous movements, they added new modules to the action repertoire. 'Legitimation stripping' in the form of uprooting crops and the subsequent entanglement with police and courts drew unprecedented media attention to the behaviour of the activists. 'Enclave building' on a local level involved engaging wider sections of civil society. Both contributed to the UK government's loss of confidence over allowing field trials to continue in 1999. However, in engaging a wider range of civil society in the GM debate the movement tended to emphasize health and consumer issues over patenting, biodiversity and global social justice as the key injustice frame of the movement.

Summary

From my fieldwork it appeared that the leadership of the anti-GM movement in Britain in the mid-1990s consisted of small networks of highly skilled activists that I have called counter-experts. This counter-expertise was multi-dimensional, including not only the technical skills and knowledge of conventional experts, but also an awareness of the social issues involved in biotechnology and an ability to articulate the irresolvable uncertainty raised by new technologies.

It was through the interplay of these three aspects that counter-experts brought technoscientific / property changes into contact with established ethical languages and crafted a new critical discourse. This discourse made power visible at the heart of a profoundly technocratic project, and opened up new political choices, a key element of new social movement activity (Melucci,

1989). The counter-experts were mediators between the private conversations of experts and wider publics. The NGOs entered the tightly bounded expert conversations as representatives of 'the public interest' and conveyed the disputed judgements made by experts out to the public. In the process the counter-experts made their own judgements, re-framing biotechnology in terms of environmental risks and human rights. It was in part the resonance of these frames with lay members of the public, which established their legitimacy as interpreters of the 'public interest'. The process of framing was an interaction between everyday life practices and inchoate perceptions of wider social groups and the sharper political projects of the counter-experts who provided a tighter focus. The counter-experts sought to recruit these social groups as their 'public' through their identities as farmers, gardeners or consumers. The consolidation of some shared outlines from the overlapping frames of the various counter-experts into a movement frame was the kind of work indicated by Melucci's (1989; 1996) challenge to dominant symbolic codes, or Jamison and Eyerman's (1991) notion of a new cosmology.

A key danger that faced the counter-experts was that in becoming competent in handling expert debates, in order to explore their inconsistencies, it became easier to communicate with their adversaries than with their supporters. Counter-experts were not the kind of grass roots organizers Jamison and Eyerman wished to find at the centre of a social movement. The NGOs involved were not participatory organizations, but professionalized and usually small. The networks that linked individual actors in these NGOs concerned with seeds, biotechnology and patenting had a deceptive double quality. The NGOs were fond of putting themselves forward in relation to government as coalitions of organizations usually including mass membership organizations such as Greenpeace. Privately, however, the NGO viewed their networks as consisting of committed and competent individuals, who did the intellectual work. More ambiguous support came from bigger organizations, which often held only a watching brief. While the organizations had their own objectives and styles of work, the networks were small and depended on personal trust and individual reputations. The UK NGOs could not be separated into a national anti-GM movement. Their concerns and connections were persistently global. Focused on the global regulation of seeds, they were part of a global movement and a global civil society. Subsequent expansion of the movement in the UK has had international impacts but may have thinned the range of issues under discussion.

5 Cosmopolitan Networking: Counter-Experts and Global Civil Society

A social movement must achieve three core tasks. First, it must build networks and develop a collective identity, usually linking together existing networks, and negotiating between competing identities. Second, it must frame a conflict and its adversary. Often this opens up a new field of conflict. Just as networks and identities in a movement are multiple, so too are the issues and adversaries. Third, a social movement must employ an action repertoire to engage its adversaries and develop a sense of agency. Social movements operate within civil society, the field of relatively free association distinct from either state or corporate economy. Civil society is not a traditional community. It is a relatively rationalized public sphere open to competition and choice as well as intimacy and solidarity. Yet civil society like social movements is no longer contained in national boundaries. Global civil society is coming into being along with global social movements. Both movements and civil society have a tangible global existence in global networks (Shaw, 1994). More intangibly global movements frame conflicts as global issues, situating participants within global civil society (Hegedus, 1989). Furthermore global movements represent civil society in and around global institutions (Shaw, 1994). In the case studies explored in this chapter and the following one, the global nature of the anti-GM movement will become clear in the global NGO networks come together to frame global issues of biodiversity and seed patenting and take collective action to engage with global governance.

This chapter presents a case study of an event in which seed activists meet to embody global networks. The first is a study of an International Symposium in Bern, where a predominantly European group of NGOs and academics met, followed by a small meeting to set up a European NGO network. The concerns of the symposium were far more global than European. Several of the keynote speakers were North American, and some from India and Africa. A cosmopolitan network was consolidating, with global links and sharing a global frame, whose object was to transform global regimes.

The anti-GM movement's identity frame had to link cosmopolitan counter-experts (mainly Northern) and local seed savers (mainly Southern) in a collective solidarity, which allowed the cosmopolitans to speak on behalf of the locals, without loss of credibility. The anti-GM movement framed the ethical issues of patenting in terms of both justice and sustainability, using interconnected 'risks' and 'rights' frames. Movement frames must be constructed in order to deal with ambivalence and produce temporary rhetorical 'solutions' to irresolvable dilemmas between contradictory discourses (Billig, 1995; Melucci, 1996). Ambivalence to the process of globalization ran through all these frames. The injustice frame warned that global corporations wished to enclose genes and knowledge of local communities, posing risks to biodiversity and infringing the rights of farmers and indigenous people. A counter-expert intervention also raised the dilemma of the ambivalent role of expert discourse in facilitating enclosure. In framing the agency of the movement, two dilemmas presented themselves. The first dilemma was how to uncover the structural causes of biodiversity loss without denying the small anti-GM movement any ability to push change in the opposite direction. The second dilemma contrasted between their desire to weaken global power structures and to strengthen local resistance to globalization, while recognizing the need to enter global governance arenas.

Establishing a collective identity requires the formation of networks that facilitate flows of information, symbolic codes, personnel and affect. More informal 'breaks' in the symposium provided time and space for more expressive ways of developing a collective identity, through communal eating, sleeping, drinking, walking, dancing and listening to music. These communal activities also facilitated making new contacts and re-affirming established ones, so that the flow of information and negotiation of positions was much more complex than in the formal presentations.

It also involves some degree of frame alignment between individuals and groups within the movement, and is therefore an ongoing process of negotiation (Snow and Benford, 1988; Melucci 1996). The formal aspects of the Bern symposium were organized to generate movement frames through the use of keynote speakers and workshops. Thus in the anti-GM movement the processes of consolidating a social movement happened at the global level. Solidarity was negotiated between globally diverse constituencies and their networks, a common discursive frame for articulating a global conflict developed, and the system of global governance confronted in the form of an inter-governmentally negotiated regulatory regime. The anti-GM movement identified its adversary globally. It recognized itself as a movement and achieved some level of recognition as a global social actor.

Patents, Genes and Butterflies in Bern

The international symposium entitled, *Patents, Genes and Butterflies*, on 20 - 21st October 1994 in Bern, was organizationally framed as a link between the concerns of the environmental movement and the development movement represented by WWF and Swissaid, respectively. Patenting of genes was to be approached as a problem of both social justice and environmental sustainability. Prominent figures in each organization were given slots in the opening sessions of the two days in which to attach their organizational labels to the proceedings, Claude Martin, Director of WWF on the first day and Gabrielle Nanchen, President of SWISSAID on the second day. This linkage of organizational trademarks obscured temporarily the political strain within WWF over what should be admitted into the canon of environmental issues.

The three individuals organizing the symposium were Miges Baumann from Swissaid, Florianne Koechlin from WWF (Switzerland) and Michel Pimbert, head of the WWF (International) Biodiversity Programme. (Florianne Koechlin was also European Co-ordinator of the No Patents on Life Campaign, and organized direct actions, while Miges Baumann and Michel Pimbert were both GRAIN Board members). Michel Pimbert had previously worked for the CGIAR system, where his community-based approach to seed research had been perceived as too radical. He had thus migrated over to the NGO sector, where he was apparently able at this point to bring the weight of WWF behind a critique of patenting. However, within a month of the symposium, Pimbert was summarily dismissed by the council of WWF, embodied by Claude Martin, the same individual who opened the symposium in Bern. The symposium linking conservation of biodiversity to community became an exemplar of the kind of work WWF had decided to avoid, and was cited as a reason for Pimbert's dismissal (*Guardian*, 18/11/95)[16].

The physical setting of a meeting always has an impact on the tenor of the meeting. The formal proceedings of the symposium took place in the Kursaal, a large lecture theatre in the shape of an amphitheatre, with fixed desks and seats banked up. About 250 people attended the plenary sessions, leaving plenty of space. The speakers and chairpeople spoke into fixed microphones, and roving mikes were used for questions or comments from the audience. Most participants were academics or NGO activists. Only one member of the audience identified himself as an industry person, and none of the speakers were drawn from industry, as is to be expected from a social

16 The proceedings of the symposium have subsequently appeared as a book, *The Life Industry: Biodiversity, People and Profits* edited by Miges Baumann, Janet Bell, Florianne Koechlin, and Michel Pimbert (Baumann et al, 1996).

movement event. All the presentations and discussions were conducted in English, German and French, with simultaneous translation. In fact, large sections of the symposium, especially the discussion, were in English, including contributions from Nordic or Dutch participants. Nevertheless the process of simultaneous translation introduced both a certain formality and a flavour of international collaboration. The workshops each contained between fifty and eighty participants, and therefore largely failed to provide a more intimate opportunity for networking, especially those in the Main Hall, which proved a little inflexible as a medium for the more intensive, network oriented discussion, due to the size of the room, the formalized physical structure, and the use of microphones and translators. In fact, these structural conditions favoured participants with set lines and prepared speeches. Human networks are entangled with technological elements and the particular form they take (international, formal) is embodied in technological ways (via buildings, seating arrangements, simultaneous translation technology). Successful operation in this milieu requires the ability to form alliances with these non-human actors, to enrol them into one's own network, as Actor Network theorists have argued (Callon, 1986; Latour, 1987).

Collective Identity and Networks

The embodying of the anti-GM movement, or indeed global civil society, in an event such as this, stretches beyond rhetorical exchanges of opinion. It has been argued that communal eating and drinking has always played an important role in diplomacy (Constantinou, 1995) as a non-logocentric form of communication separate from the negotiation of interests. More generally, the sharing of food (and food taboos) provides a sort of substratum for developing trust and thus embodying a community as a community (Eder, 1996a) alongside the struggle for power via negotiation. The expressive dimension is never absent from social movement formation, even as in this example where the instrumental work of dedicated counter-experts is the staple fare of the movement activity. This less apparent expressive sociability (Simmel, 1971), sociality (Maffesoli, 1996), or 'Dionysian principle' (Nietzsche, 1956), with its emphasis on music, pleasure and intoxication, can be as important for social movements as its opposing 'Appollonian principle', concerned with vision, dreams, individualization and Statecraft (Nietzsche, 1956). Certainly attempts to classify social movements as either instrumental or expressive (Kriesi *et al*, 1995) require substantial revision.

Thus we might reverse the usual order of priority and suggest that the formal meeting in Bern was secondary to the informal networking that took place in the coffee breaks and meals. It was then that a multiplicity of one to one or small group contacts were made. To the actual process of consuming food, must be added the chance meetings in the toilets (the gents at least seemed to be a meeting point where old acquaintances were greeted and key points discussed). A particularly valuable unstructured space for discussion occurred during the long lunchtime walks across the bridge that spans the gorge separating the symposium venue from the nearest restaurant.

A lot of the most interesting material for me appeared through conversations of varying lengths in these breaks in the proceedings. Robin Jenkins of Genetics Forum, whom I met in the lobby before the start of the first session, remained my most important contact. He provided a commentary not only on the symposium itself, but also entry into the GENET NGO meeting that was organized on the day following the symposium to piggy-back on the presence of so many NGOs gathered in one place. He also spoke on a wide range of issues, including forms of networking and the immanent National Consensus Conference on Biotechnology in London (See Purdue, 1999). It was in these informal times that I could meet a number of academics and NGOs, and aspirant NGOs[17]. My own activity in these breaks was unexceptional. This sort of networking can take several forms: the pursuit of those it is strategic to connect with, such as Pat Mooney or Vandana Shiva, or key members of the press; catching up with old friends; random connections with new people who may be interesting, or attractive. I, like other delegates, used a combination of all of these forms.

Existing networks came into play. Allocation of accommodation contributed to the structuring of social interaction. Key figures, who were paid to attend, were grouped together in a single hotel, together with some ambitious hangers on. One such hanger on confided that the vast amount of money it had cost was worthwhile to be in the same place as Shiva. The Youth Hostel, on the other hand, contained many of the PhD students and representatives of small NGOs. Initially, the enforced intimacy of Youth Hostel domestic arrangements generated hostility towards each other. As the symposium progressed, hostility gave way to a sense of common purpose, as the other occupants of the Hostel became recognizable as fellow delegates, leading towards communal breakfasts and other informal discussions, even in the rather Spartan surroundings.

---

17 These included Dutch academics, working with Southern NGOs to set up local seed banks, a Norwegian political scientist, analysing the Biodiversity Convention in terms of regime theory, the Guardian's environmental editor, who thought the NGOs had lost their way after Rio, a member of a Swiss NGO which is organizing a referendum on release of GMOs (they already had the requisite 100,000 signatures), and Janet Bell, one of the editors of the symposium book, who had previously organized the Gene Traders Conference in London.

Most participants were obviously in Bern to meet other people with similar interests and made use of the opportunities for socialising, including the provision of a Ghanaian band, Akapoma, on the final evening. Dancing together allowed an ecstatic sense of community to surpass strategic calculations, at least temporarily, as the basis of a shared identity. As well as the acknowledged danceability of African music, it underscored the symposium's commitment to a politics of support for the South, even though very few of the participants came from Southern countries. The music played an expressive part in the symposium's political opposition to a perceived racism in the gene industry's treatment of indigenous peoples and Southern farmers, displaying a symbiosis between the Dionysian sociality of the music and the strategic framing of the anti-GM movement.

Global Identities: Cosmopolitans and Locals

Tewolde Egziabher (1994), a scientist and spokesperson for Ethiopia and the G77 countries in the FAO negotiations voiced the dilemma of an identity for a global movement, lying across the North-South divide. He spoke of the need for people of the North, whom he called people of the biosphere, to make links with those in the South, whom he referred to as eco-system people. Such biospheric reach made Northern NGOs valuable allies to Southern people, who are bound by their dependence on their immediate eco-system.

This formulation of two contrasting subject positions within global civil society of free floating, universal agents in the North, and rooted, embodied people in the South, connects with debates over the simultaneous emergence of modern science and modern civil society in Europe. Both it is argued depended on the disembodied white, male, bourgeois as a civil, modest witness to reflect the veracity of scientific knowledge claims. By contrast, women, black people, the working class, and so on, were portrayed as refracting the scientific gaze through the 'special interests' they embody and therefore remain marginal to both science and civil society (Haraway, 1997).

The radical difference between Northern and Southern identities, given an ecological grounding by Tewolde Egziabher, leaves global civil society as a Northern civil society. People of the South appear only by proxy, through their supporters in the North. The North-South imbalance in global civil society also explains the presence of representatives of Southern governments or their advisors, alongside Northern NGOs and individuals, within the anti-GM movement and therefore within global civil society.

Nevertheless a deep sense of ambivalence ran through counterpoising the local and the global throughout the symposium. A general rallying call was for support of 'local communities' particularly in the South, supplemented at one point by the claim that *'we are all local people'*. This ambivalence may have arisen from the likelihood that most of the participants have little or no contact with a local community in any traditional sense. Rather than members of traditional face to face communities, directly affected by loss of biodiversity or loss of intellectual property rights over their local communal resources, the conference appeared to attract knowledge-intensive social actors who occupied a global space. Their primary identity, vis-à-vis biodiversity, was as members of social movement networks or global civil society, linked largely through telephone wires, print and cyberspace. Hence the importance of this event, where the cosmopolitan counter-expert networks can be temporarily embodied in a face to face community. Just as expert systems are re-embedded in everyday life through face-work (Giddens, 1990).

Support for cultural diversity depends on a symbiotic relationship between cosmopolitan and local cultures. Cosmopolitans value diversity as such, whereas locals require diversity between different locales, to protect their locality and way of life from homogenization (Hannerz, 1990). Given the established argument that biodiversity depends on cultural diversity, it is a small step to claim that NGOs may be viewed as cosmopolitan allies of local food growers, sharing a common interest in protecting biological and cultural diversity, but from very different cultural standpoints. Biodiversity is an immediate material resource issue for Southern farmers, an insurance against capricious nature. For the Northern NGOs and seed savers, it may be seen as embodying post-material values (Inglehart, 1981; Dyer, 1993).

The identity frame had a significant weakness. Those present fell into the cosmopolitan category, while rhetorically invoking absent 'local communities of the South' as the root and power base of the movement. Vandana Shiva attempted to resolve the question of who the 'we' of the anti-GM movement was by mobilizing her own multiple identities *'woman and scientist, farmer and woman'* to indicate the diversity of the movement (Shiva, 1994). Only Charles Zerner reminded the symposium of the unstable ethnoscapes of the global era, directing attention to migrants and marginals, for whom the notion of a stable local community is ephemeral at best. Little clarity existed as to how seeds were being saved and distributed, by and between individuals, families and communities.

> *In this era of intense mobility, social uprooting and rapidly migrating capital, it is becoming increasingly difficult to specify, in simple social and topographic terms, the boundaries of particular communities. In an era where plants, peoples, ideas and molecules travel and combine in unexpected, and often unpredictable, ways.* (Zerner, 1994)

The Injustice of Patenting: Framing the Issues

The shape of the agenda for the meeting and the kinds of alliances the organizers were hoping to establish were signalled in the subtitle of the symposium **'Are plants and Indians becoming raw materials for the gene industry?'** The first day of the symposium was given over to developing a critique, whereas the second day aimed to develop alternatives. Yet, far from giving a clear set of alternatives to the dominant practices, speakers mixed proposals, critiques and rhetoric. The Human Genome Project was linked to racism, by Alan Goodman, while the Human Genome Diversity Project was somewhat more gently dealt with in terms of a methodological critique by geneticist, Andre Langaneyan, an ex-Committee member of HGDP. Bio-prospecting and the Body Shop model of marketing non-timber forest products were explored as attempts at equitable models of co-operation with local communities. Vandana Shiva and Robert Chambers attempted a more general critique of the dominant development paradigm.

As the first substantive speaker, Jack Kloppenburg undertook to shape an injustice frame. He began by disentangling hegemonic dominance from inevitability, or progress. He asserted that patenting was the outcome of a history of conflict. Therefore, it was necessary to take sides. He defined the anti-GM movement's adversaries in a long sequence of oppositions.

> *I am NOT on the side of rBGH, I am NOT on the side of Monsanto, I am NOT on the side of NAFTA, I am NOT on the side of GATT, I am NOT on the side of WTO, I am NOT on the side of those who want to see this wonderfully diverse world flattened biologically and socially to accommodate the unrestricted flow of capital and commodities. I am not on the side of those who would make plants and American Indians raw material for the gene industry. I am not on the side of those who want to continue to cut up the world and parcel it out for sale.* (Kloppenburg, 1994).

The sides one was encouraged to take were for or against enclosure / commodification, with biotechnology framed as an imperialist enclosure project.

> *This process must be seen as part of the larger historical process of global commodification in which nothing is sacred and anything can be bought and sold.* (Kloppenburg, 1994).

The history of biotechnology has been characterized by six constants

- *Northern states and the military underpin patents.*
- *The West sees the world as its garden, to collect and change, at will.*
- *Genetic and cultural information are always collected together.*
- *The North owes a massive genetic debt to the South in terms of the labour invested by Southern farmers in developing seeds and knowledge of plant uses.*
- *What was Common Heritage in the South, is turned into commodities in the North.*
- *There are two, or more, sides to these issues.* (Kloppenburg, 1994).

When Kloppenburg suggested that *'there are more than two sides, but to recognize that there are two is a start,'* he clearly put the emphasis on the second part of this formulation, which quickly fell into a them-and-us pattern, the big Northern TNCs and states versus the people.

Kloppenburg explicitly sharpened the 'hot cognition' of injustice into an 'adversarial identity frame' (Gamson, 1995). The anti-GM movement was able and willing to name its adversaries in the globalising North. However, it was better at defining 'them' than defining 'us'. Framing issues may also involve the three Kantian moments of matters of empirical or scientific fact, moral issues and aesthetic judgements (Eder, 1996b). A risk frame is intended to focus attention on the uncertainty of the empirical facts of the outcomes of technological change, and criticises the scientific methods intended to claim certainty in these empirical outcomes. A rights frame deploys a moral critique of the distributive equity arising from the introduction of new technologies. Aesthetic judgement too is involved if a critique is to use terms like the biological and social 'flattening' of 'this wonderfully diverse world'. Later speakers such as Vandana Shiva amplified this position with a call to arms to oppose 'the enclosure of the interior spaces of our bodies'. All the diverse constituencies and identities affected by patenting could, she felt, be mobilized around the slogan 'No Patents on Life'. The presentations, in general, offered a fairly simplistic and rhetorical call to oppose the patenting of life, genetic determinacy and current development practices.

The enclosure frame outlined by Jack Kloppenburg was blended with a critique of expertise, as the plenary sessions on the first morning continued, with two closely argued contributions on the contradictions within the expert discourses of genetic patenting. Christine Noiville outlined the differences between European and American Intellectual Property Law. Regina Kollek drew attention to the difficulty in defining genes.

Noiville (1994) focuses attention on the clauses of the European Patent Convention, which allow for moral and ecological objections to particular patents (Clauses 53a and 53b). She raised the question of whether Patent Law, as a legal system, could deal with these issues, or whether they could best be dealt with elsewhere. This is a generic problem of counter-expert politics. Expert systems are designed to foreground technical matters and leave broader political implications in the shadows (Haraway, 1997). When counter-experts intervene, and translate social issues into the language of expertise, they put these expert systems under tension by challenging the underlying assumptions, but nevertheless are themselves constrained by the expert discourse. So counter-experts attempt to read loophole clauses, such as 53a and 53b, or the *sui generis* provision in the TRIPs, against their intended function.

The presentation entitled 'Genes: A Diffuse Object of Desire?' (Kolleck, 1994) opened up to investigation the stable 'Blackbox' definition of genes as discreet entities, with clearly defined functions, essential to patent law. Kolleck suggested instead, that there are multiple definitions of genes at use within scientific work, and also multiple theories of the link between genes and the expression of particular traits. There is no consensus that shifting a particular piece of genetic material from an organism of one species to an organism of a quite different species, would produce a stable transfer of the trait possessed by the first to the second. However, these sorts of knowledge claims are essential to the patenting of genes.

By lunchtime on the first day, two types of criticism of the format had emerged. On the one hand, an academic expressed reservations over the predominance of rhetoric over more measured analysis or practical detail. On the other hand, an NGO activist, regretted the lack of sharper debates. He criticised the academic speakers for failing to follow through the political implications of their analysis. And further commented that big conferences stick to consensus-building rhetoric and smaller meetings tend to be confined to bureaucratic details, leaving no space in between for political debate. These comments were to prove an accurate assessment of the comparison between the symposium and the GENET NGO meeting that followed.

Global Agency and Counter-Expert Action Repertoire

Pat Mooney was given the final plenary session to motivate the participants for action. Rather than specify the elements of the action repertoire, he attempted to provide an agency frame (Gamson, 1995) which could give a credible purpose to the action undertaken by the movement within a global political field. Mooney was introduced as *'the first in the field'*. He celebrated the growth in the size of the movement '*15 years ago this conference could have been held in a telephone booth*' (Mooney, 1994).

His address developed the butterfly as a metaphor for agency. The butterfly effect in chaos theory suggests small events can have massive results, therefore huge global institutions such as the WTO could be vulnerable to the action of small social actors. Hence the possibility that a small movement may have been able to make a significant intervention in the play of global forces. This formulation spoke to a key problem of agency in the anti-GM movement. The movement was up against very powerful global forces, and could count on very few active members, whilst claiming to speak on behalf of a much wider constituency of followers, with whom their relationship was entirely unclear. Hence the political work that Mooney undertook with the butterfly metaphor was, against the odds, to give the anti-GM movement a credible sense of agency - a belief in itself as a subject of social change on a global scale. NGO butterflies, he argued, were able to rise above dinosaurs, such as the WTO. This represented a novel 'solution' to the enduring dilemma of exposing, or naming, the structural power of the hegemonic adversary, while simultaneously developing the sense of agency in the movement's ability to challenge that hegemony. In more linear times, Gramsci (1971) called this dilemma the pessimism of the intellect and the optimism of the will.

The idea of seeing social movements as butterflies had other latent advantages. The ephemeral, the sporadic, the cyclical, the mobile could be turned round from criticisms of environmental action to proof of an appropriately non-linear kind of politics for a turbulent age of global 'chaos' (Rosenau, 1990). Pat Mooney suggested that, as butterflies flit about, so should NGOs shift back and forth, between the local and the global. A second dilemma of agency was indicated: how to confront the power of global adversaries, while defending locality as a credible arena of action. While it is easy to give the injunction to 'think globally, and act locally' other permutations of these terms come more readily to hand. The loss of biodiversity by local communities needed to be confronted at a global level, and yet simultaneously, opposition to patenting involved a defence of the local against globalising power. The strengths of the NGO community lay in its global networks and counter-expert skills, not in local rootedness. However, an ability to connect with local communities was an essential ingredient in their advantage over their adversaries as agents of legitimacy. Mooney advocated not only that these global travellers must touch ground somewhere, but that the rooted locals must appear at a global level, which in itself constitutes an erosion of locality.

Throughout the symposium the local was repeatedly invoked rhetorically as the power base of the movement, whilst the global was treated with suspicion as enemy territory. Thus the local-global divide was utilised, as one element of the contrast between the movement and its adversary: global corporation and global regimes against local people or local communities. Action at the global level remained a subject of intense ambivalence. Whilst Kloppenburg proposed that '*the Biodiversity Convention is an empty document ... Our record at the global level isn't good*'. Pat Mooney opposed him with the claim that '*the Biodiversity Convention is a weak document that can be made into a stronger one'*. Global action was required. Rather than a simple opposition between a local 'we' and a global 'them' the appropriate level of action for the movement remained an unresolved dilemma. Pat Mooney's butterfly metaphor provided a creative 'solution' lacking in the position of committed localists, whose paeans to the local commons sit uncomfortably with the global nature of their campaigning work and its environmental infrastructure.

Summary

In summary, the symposium acted as a nodal event in establishing the collective identity of the anti-GM movement, linking Europeans in a global context. Charismatic leaders such as Jack Kloppenburg, Vandana Shiva and Pat Mooney, were drawn in by the organizers to frame the movement in a way that would bring together as wide a range of constituencies as possible. Probably more significant than the formal network building was the complementary process of informal networking, that is important in developing a density of networks with multiple connections essential to a social movement with no centralised structure. The international symposium in Bern was a moment in which the anti-GM movement constituted itself as a set of counter-expert networks within global civil society. Leading actors within the anti-GM movement attempted to frame the issues of concern, the collective identity and agency of the movement at a global level. The leadership also facilitated more expressive and spontaneous networking, which established numerous cross linkages between individuals. In the FAO meeting in Leipzig, these processes continued, but were extended by their representation of global civil society in relation to the inter-governmental negotiation of a global seed regime. Seed experts and counter-experts face one another directly at the FAO, which had long been a favoured site of global seed politics.

6 Global Civil Society and Global Governance: NGOs and the FAO in Leipzig

This chapter consists of a second case study, which concentrates on NGO activity representing civil society in the FAO, a global institution forming part of the regime regulating global gene flows. In the process of mobilizing around the FAO, NGOs also consolidated their global civil society networks to strengthen the global anti-GM movement.

The Food and Agriculture Organization (FAO) was set up in 1945 as a branch of the UN with specific concern with food and agriculture globally. Agro-biodiversity became a particular concern of the FAO with major conferences on Plant Genetic Resources in 1961 and 1967. The FAO was the site of the 'seed wars' of the early 1980s when Southern countries argued that seeds were a global commons and that even those seeds held in private seed company collections should be generally accessible (Kloppenburg, 1988a; 1988b). The FAO passed an International Undertaking on Plant Genetic Resources in 1983, which declared seeds to be a common heritage of humankind and formed the Commission on Plant Genetic Resources. In June 1996, the FAO's Commission on Plant Genetic Resources held its Fourth Technical Conference in Leipzig. This meeting was to pass a new International Undertaking on Farmers' Rights to become a protocol to the Convention on Biological Diversity. It was held to be one of an important sequence of international meetings, linking the concerns of food security and agro-biodiversity in the FAO with biodiversity in the CBD and international trade in the WTO. The linchpin integrating these global regimes was to be IPRs over plant genetic resources. That is, who owns the seeds? NGOs identified two contending positions: a global corporate agenda of patenting versus a global civil society agenda of Farmers' Rights.

The Fourth Technical Conference of the PGR Commission was held in Leipzig on 17-23 June 1996, after a gap of 13 years since the 3rd Technical Conference in 1983. The build-up to Leipzig included each member country preparing a report on the state of its plant genetic resources for food and agriculture (PGRFA), which were collated through sub-regional and regional meetings into a Global Report on the State of the World's Plant Genetic Resources. On the basis of this report members of the FAO secretariat wrote a Global Plan of Action and a draft of the Leipzig Declaration. These two documents were to be agreed in Leipzig and to lead to a new International Undertaking embodying Farmers' Rights to be adopted and forwarded to the Commission on Biological Diversity as a protocol of the Treaty on Biodiversity. Much of the text of the global plan of action and the declaration was challenged by various countries, largely the USA and Canada, and thus put in square brackets to demonstrate lack of consensus. A meeting of the PGRC in Rome in April resolved many of the differences on the global plan, leaving the more difficult square brackets for Leipzig, but the Undertaking was deferred to a later meeting in December. Thus Leipzig should be viewed not as a discreet entity but part of a longer process, in which the decisions taken in Leipzig were tied to precedents from previous meetings or deferred for further interpretation at future meetings.

Global Governance in Action

Hegemony and Marginalized Inclusion

Every level of the Leipzig process displayed a key mechanism of hegemony - the marginalized inclusion of oppositional discourse. Entry into a hierarchical system is never complete, always conditional. Sites of hegemonic power should not be reified in critical discourse, as the power may shift elsewhere, leaving a newly occupied site without the symbolic charge for which it was originally sought. Thus it has been argued that the FAO itself has been marginalized as a global institution regulating IPRs, as corporate power shifted to the GATT / WTO. Nevertheless, NGOs have used it as a spring board, from which to attack other global institutions such as the CBD and WTO. The FAO does, however, share the NGOs' core concerns with food security, agriculture and agro-biodiversity. Thus a major struggle for the seed activists was to overcome their marginalization at the conference and make their inclusion 'real'; in other words, to make their presence felt.

The Conference itself occupied a position of marginalized inclusion within the power structure of the FAO. The new Director-General of the FAO had thrown his weight behind the World Food Summit, which was scheduled to be held November 1996 in Rome, as a project of personal glory. The Summit was intended to advocate industrial agriculture and reliance on the global market for food security, and so although it was to have no decision-making remit, it had massive resources dedicated to it. Leaked memos revealed that TNCs were offered an agenda point each, in exchange for a contribution of $1 million. By contrast, preparation for the Leipzig Conference, including the work of writing the global plan and other documentation, fell to a small secretariat. NGOs, Cary Fowler from RAFI and David Cooper from GRAIN, were drafted in to supply the expertise, but little or no secretarial and other resources were allocated[18]. Yet Fowler and Cooper were employed on fixed term contracts that expired four days after the end of the conference, leaving only two professional staff to organise the follow-up meeting in December.

The same process of marginalized inclusion was inscribed in the presence of the NGOs in the Conference site. The Conference centre was part of a newly built complex, on the outskirts of Leipzig, with its own tram stop at the end of the line. The complex was a symbol of West German investment in the newly integrated East. The complex included a massive glass dome, dubbed 'Biosphere 3' by NGOs. For some NGOs the technocratic building and the institution of the FAO were barely separable.

Initially the NGOs had been denied access to any of the 12 halls and seminar rooms in the conference centre, to hold meetings. Therefore the NGO committee hired a tent for DM 25,000 (approx. £12,000), which was set up in a car park, behind the building. The air-conditioned building, with its coffee bars, open atrium and high-tech translation technology was separated from the NGO tent by a long walk around the building and through a car park surrounded by a 10 foot security fence. Thus very few governmental delegates ever found their way to the NGO tent. Had it been located in front of the building en route from the tram and bus stops, its symbolic role may have been enhanced. The gate in the security fence became an obligatory point of passage for anyone attempting to gain access to the NGO tent. As neither the conference centre nor the FAO was prepared to pay security personnel, the NGOs had to supply a rota to staff the gate - the least enviable of the tasks in the hierarchy of the global conference. Thus gate security usually fell to the teenage daughter of one of the German NGOs, or sometimes members of the local Leipzig environmental group, Eco Leeuwe[19].

The first NGO press conference was held in the tent, in great heat. The presentations were almost drowned out by the noise of earth moving equipment Working on the road outside. The smell of sewage came later in the week, when the tent was being used only for eating the communal NGO lunch, while formal meetings had been relocated into the building. A German filmmaker summarized NGO feelings well, when he commented *'It's really like*

---

18 Cary Fowler, who headed the secretariat, reputedly had to answer phone enquiries from ambassadors wishing to bring their wives, while he was trying to write the final drafts of the documents to be discussed.

19 Symbolic marginalization can take a very physical form. One evening I missed the opening times and had to climb the fence, using a massive rubbish bin as an escape route.

apartheid'. When NGOs abandoned the gate rota, and started using the kitchen as a route to their tent, they were officially reprimanded, drawing a sarcastic response from a German academic *'We must use the proper servants' entrance'.*

As much as anything else this marginalization appeared to be the product of inflexibility at all levels from the physical structure of the building and the related insurance and safety regulations, through to the inflexibility of the conference organisers in dealing with allocation of rooms. In spite of the surfeit of rooms, NGO meetings were still frequently relocated to avoid clashes. How much of this inflexibility was rooted in the monomania of global hegemony and how much simply reflected the national and regional cultures of Germany and former Eastern Europe remains a moot point. The NGO office was at least within the building, although it was located at an extreme point, and difficult to find. The e-mail never did work, and getting access to enough functioning photocopiers was an ongoing battle. Exhibition space was similarly organized, so that the NGOs had the most distant stalls.

Compared to the treatment of NGOs in Beijing or even Rio, these may seem small inconveniences. The cumulative effect, however, was to create a sense that few areas of conference life lay outside the inscription of status hierarchy. To those at the centre of power, business simply continued as usual. For the NGOs and student activists, their marginal status was continually re-enforced by their experience of the details of conference life. The instrumental logics of security, environmental health and distribution of resources were invisible to the former group, but a continual constraint to the latter. These instrumental frames may have been intended simply to bracket off the day to day running of the conference centre from the work of governmental delegates. Yet they subtly dramatised a deep cultural clash between the ways of life of the bureaucratic diplomats and the 'protesting' social movements (Weber, 1948; Moscovici, 1993). The collective and relatively egalitarian social movement culture fits poorly in the hierarchical ordering of social life by status and predictability. The counter-experts sat on a knife-edge, always balancing between inclusion and marginalization. On the one hand, they were drawn towards inclusion through the most expert presentation of their intellectual critiques of current practice in an attempt to win the arguments. On the other hand, their marginality was inscribed in every aspect of their life at the conference. By embracing aspects of a social movement lifestyle, they expressed a cultural difference and made visible the power frames regulating acceptable forms of social life.

The meeting itself contained internal procedures of marginalized inclusion. The NGOs had observer status and were allowed access to the plenary sessions, where they were allowed to speak at the discretion of the chair but only at the end of an agenda item. Only seven NGO interventions were made in total; less than 30 minutes in the seven-day conference.

Furthermore, the real decisions were made in closed working groups, the most crucial of these were ad hoc groups, which went by the name of Friends of Chair, where the Americans and Canadians worked out their differences with the rest of the world[20]. Information as to what was going on behind closed doors was not totally blacked out. Several individuals were both governmental delegates and NGOs, including the head of the Ethiopian delegation, Tewolde Egziabher, who played a key role in the African and G77 blocks. It was obviously the skilled, experienced and well connected NGOs, primarily Pat Mooney (and probably Henck Hobbelink), who were able to keep in contact with the decision making process, while many NGOs had no access except through Pat Mooney et al. At key moments many of the delegates were themselves totally marginalized. The plenary was abandoned for twelve hours on the final day, while the Americans wrung concessions from the G77 behind closed doors. As they waited, most of the African delegates grouped around the TV monitors watching Holland play France in the Euro '96 quarterfinal, along with the Dutch delegation. The occasional whoops and cheers issuing from this caucus had little to do with the Global Plan of Action on Plant Genetic Resources.

Priests and their Holy Texts: Life in Square Brackets

Global leadership in the conflict over agricultural biodiversity followed the model of leadership cycles developed from the history of religion. Rationalising priests maintain routinized hierarchies and draw their influence from established rules and traditions. Periodically innovative prophets arise from outside the hierarchy, relying only on charisma and argument; they condemn the existing system, establish new values and inspire followers to found new protesting movements. With time, everyday routines and responsibilities reassert themselves, requiring more rational or traditional forms of leadership and legitimacy, as prophetic charisma fades, power reverts to a new clutch of priests (Weber, 1948; Moscovici, 1993). Cultural innovation depends on these prophets and protesting minorities, while maintaining social conformity is the territory of the bureaucratic priests (Moscovici, 1993). In this case, the priestly class pouring over their holy texts were the diplomats representing the national governments, while the NGOs played the role of secular prophets (Melucci, 1996) declaiming authority at the city gates.

---

[20] According to a Filipino delegate, the closed doors allowed the Americans to indulge in intimidatory behaviour.

The official FAO meeting consisted of a six day long editorial meeting with the task of coming up with two agreed texts: the Global Plan of Action and the Leipzig Declaration. The editorial committee took the unwieldy form of delegations from 160 countries, working in eight languages simultaneously, in various combinations of full plenary sessions, drafting groups, working groups and 'friends of the chairman' log-jam breaking groups. What stood in the way of consensus were stark contrasts in agricultural policy expressed in punctuation terms as square brackets. A considerable number of bits of the texts were under dispute and had been put in square brackets to denote the divergence of international opinion. In general two editorial possibilities offered themselves to the 'global community' to remove the square brackets and allow the text to stand or to remove the bracketed text altogether. In general the Southern countries favoured the former change in punctuation (leaving aside various mavericks, e.g. Brazil). The US, Canada *et al* preferred the latter policy of deleting text wherever possible. An intermediary compromise often consisted of changing the disputed wording and then allowing the edited version to be admitted to the final text.

Discussions, at least in the plenary and drafting sessions, which were open to NGO observers such as myself, were dominated by two obsessions: the diplomatic and the textual. Diplomatic protocol tended to clog the workings of the meeting. Delegates would feel the necessity to represent their countries on points on which they had nothing new to contribute. Thus on any point a string of contributions would take the form of *'thank you Mr Chairman for the work you have done in hosting this meeting and your hospitality in bringing us to this beautiful city of Leipzig,'* at length followed by a quick *'we agree with the delegate from...'*

The positions taken by delegates on occasions reflected their national interests at a level of purely diplomatic prominence. Thus Germany, as host, was judged to be deeply concerned that some sort of Global Plan was agreed and that a Leipzig Declaration saw the light of day as a vehicle for their diplomatic success. According to the Maltese delegate, other countries would be unwilling to humiliate their German hosts by failing to agree a text. All through the week an agreement of some sort was anticipated, allowing the Americans and Canadians to use this general goodwill to hold out to the end for the concessions they required. Against this grain, Argentina was not particularly keen to see too much achieved in Leipzig, preferring the main action to be left to the next major meeting of the Commission on Biodiversity, in Buenos Aires, for which Argentina as host stood to gain kudos. A major component of the diplomatic form was the continual reference to a spirit of consensus. The chairman would call for the delegates to respect the spirit of consensus and their mutual desire to reach agreement, while each delegate would point out that they were in fact leaning over backwards to achieve a consensus (even in the extremely unconvincing case of the USA).

The delegates' textual obsession gave a new meaning to the phrase 'nothing exists outside of the text'. Not only was there agreement that the sole purpose of the meeting was to produce a text, but that text should be consistent with previous texts. Thus at one point a delegate suggested that we need not discuss what we mean by food and agriculture, not because these terms are commonly used and understood, but because they were already defined in the constitution of the FAO. Thus every attempt was made to pin any new term down to a definition in a previous text, with any concept that could not be so pinned down treated as a hostage to fate. The usual mantra of the protection of IPRs as defined by the WTO was woven into the text.

> Access and transfer of technology should be provided on terms which recognize and are consistent with the adequate and effective protection of intellectual property rights. (FAO, 1996a, Global Plan of Action, ITCPRG/96/5/Rev 2 - Add. 3, Para 258 (i); repeated in the FAO, 1996b, Leipzig Declaration, ITCPGR/96/6 - Rev. 2, para 7).

Which texts are or are not admissible in the global cannon or global intertext (Der Derian 1992) is of course significant. However, new terms were also left open to future battles over interpretation at subsequent meetings, particularly concerning funding and implementation. Therefore both previous and future texts are mobilised in the struggles to control meaning. This diplomatic search for a consistent web of texts securely anchored to each other's definitions and procedures, but loosely enough woven to let the bodies of starving peasants fall unnoticed, can only be likened to fiddling while Rome burns.

The NGOs pitied and parodied the high priests of diplomacy. The more pitying view described the delegates' disembodied existence as *'life in square brackets'.* They felt that that the delegates moved from airport to hotel to conference centre and back, hardly knowing where they were. They were given huge piles of papers to read, which they barely understood, and were frequently instructed what to say or how to vote by others at home. Then they were moved on to another conference on a different topic, envying the NGOs their freedom to move about, speak their minds and get to grips with the issues. The parody radiated outwards through the NGO community from a Latin American NGO, who spoke no English at all, except a frequently intoned *'Thank you, Mr Chairman...'*

Organic Prophets in Global Civil Society

Solidarity Networks and Collective Identity

The prophetic NGOs, who came to denounce and innovate, had their experience of Leipzig broken into two parts. First was a preparatory NGO meeting, entitled *In Safe Hands: Local Communities Conserving Biodiversity*. Second came the lobbying and fringe activity at the FAO itself. A third smaller element was a parallel set of seminars organized by the youth / student activist network, A SEED. *In Safe Hands* was intended as a consensus building exercise drawing in a much wider range of NGOs than ever before, including leaders from Southern peasant and indigenous peoples' organizations. Relatively absent from any of this were the Northern organic farmers and gardeners organizations[21].

As ever, the conference accommodation was marked by the ordering of wealth and power. The delegates stayed in hotels, as did some of the well-placed fixers of the formal sector. The NGOs and students were all located in a single campsite. Here too there were graduations. The NGOs stayed in huts and the students in tents. More subtle financial graduations determined the size and shape of the huts the NGOs inhabited.

Food provided a unifying focus for the movement. A Dutch organic food outfit, who usually work festivals, were employed to provide communal meals. The open-air breakfasts in the campsite became the venue for daily NGO strategy meetings. Part of the deal for the students was to do the day's washing up at breakfast time, making attendance at the strategy meetings a little difficult. This clash of roles added to a sense that the students were not actually NGOs at all, but some other political beings, and therefore aloof from the NGO strategy. Opting out of communal meals also allowed for more selective networking in the campsite restaurant, again denied the students, for whom the economic imperatives were sharper.

The opening evening in the campsite was the occasion for activists from different countries to re-acquaint themselves, called the gathering of the clans by one particularly keen networker. As I have argued in chapter eight, the shared identity of strategic counter-expertise does not exhaust the networks.

21 The Soil Association, the UK's largest organic organization, had agreed a last minute request of mine to represent them, but in the end I was on the IT delegation. While I appreciate the Soil Association's cooperation, it also demonstrates their lack of investment in the issue. HDRA, which has after all a strong presence on seed conservation and had hosted one of the preparatory meetings, was not represented at all. Another NGO said he would rather spend his time lobbying Sainsburys than the FAO.

These networks also involved exchanges of affect. The intermittent meetings do not diminish, and possibly heighten, the intense or ecstatic nature of the community. Once again, this is a cosmopolitan network in which competence in other cultures is crucial (Hannerz, 1990). Due to the large Latin American contingent, Spanish functioned as a second language to English[22], for the NGO community. Competence in Spanish became an exceptionally important cosmopolitan skill, complemented by travel in Latin America. Exchanges between NGOs frequently drew on knowledge and past experience of living in each other's countries. Desire for the Other comes in many forms: from mild curiosity, through interest in friendship, sharing each other's food, drink, music and dance, to sexual attraction, flirtation, and love affairs. The gathering of the NGO clans clearly encompassed most of these pleasures.

Framing Conflict and Building Consensus

Following a rather tense time in Rome at the preparatory meeting in April, during which a rift opened between the Latin American NGOs and the rest, the NGOs directed their energies towards building a consensus. The NGO meeting was held in Leipzig's New Council House, a rather grand building with interpreters' booths in the plenary room. All sessions, including the workshops in smaller rooms, had simultaneous Spanish translation. In the strategy meetings key participants would sometimes switch between chairing, making key contributions and translating. The presence of the translation headphones, a scattering of laptops, and the constant ringing of mobile phones, lent a relentlessly hi-tech feel to the proceedings.

The brief opening plenary session was addressed by Caninawa, an octogenarian peasant leader from the Philippines; Professor Naranswami, leader of the 10 million strong Karnataka Farmers Union in South India, was also invited to speak, but preferred to remain in the audience. Nevertheless his presence together with Caninawa, indicated the potential mass base of the anti-GM movement globally. Key anti-GM movement figures contributed some of the threads tying together a global movement identity through their personal narratives. Vandana Shiva, of India and Professor Worede of Ethiopia, both introduced themselves via their route into their present concern with seeds. In each case this consisted of a genealogy of who influenced them in their change of heart, with Pat Mooney inevitably featuring as a catalytic influence. These webbed accounts, with their mutual acknowledgements of esteem and influence, provided a complex genealogy linking key anti-GM movement actors from different continents into a global and cosmopolitan identity.

---

[22] German was used to a much lesser degree.

In contrast to the *Patents, Genes and Butterflies* meeting in Bern the prime means of framing the meeting was not through major theoretical contributions in plenary, but rather more methodological in setting the format (workshop discussion) and goals for the meeting (production of a Peoples Plan of Action). Thus, in a rather low key introduction, Henck Hobbelink of GRAIN, one of the architects of the consensus-building frame, suggested that participants should refrain from sitting with their laptops, writing position papers in the sessions. Instead, he suggested most of the meeting would consist of workshops, where an exchange of views should lead to building a consensus. This plea did not stop one Latin American NGO from reading a lengthy statement from his laptop that had been prepared at an earlier Latin American meeting. Nevertheless the process of discussion in topic workshops, and overlapping regional workshops, produced a lot material to be woven into a Peoples' Plan of Action. It also allowed some development of trust in small groups, to provide cross cutting ties, linking the various constituencies together.

In theory, the first day was spent on broader discussions and the second on producing proposals for the Peoples' Plan of Action. This Peoples' Plan, and indeed the whole meeting, was framed by Henck Hobbelink, as relatively autonomous from the FAO. That is, participants were encouraged to build their own Plan of Action, not to simply respond to the FAO agenda. A drafting group (Henck Hobbelink and Kristin Dawkins) then wrote up the Plan overnight. While most of the NGOs partied on Saturday night, Hobbelink appeared for the first time at 3am, the draft completed. The more peripheral NGOs and students found their more expressive ways of bonding. So while the leaders drafted others danced to each other's music.

The third day was spent refining the draft, with detailed objections made. This format allowed those who felt a strong need to be heard to speak a number of times. The final product seemed to be accepted as a consensus document. The People's Plan of Action was viewed as an agreed position from which to move forward after the whole FAO conference was over. This process of strategic frame alignment (Gamson, 1995; Snow and Benford, 1988) was crucial to collective identification.

Organization and Leadership

The organization of the NGOs resembled an inverted pyramid. At its base was the tiny local Leipzig green group, Eco Leeuwe, which had to take care of the basic day to day running of the NGO campsite and supplying food. Hence they spent a lot of time driving back and forth between the campsite and the conference venues bringing food and serving it, doing duty on the security gate. The next level was occupied by the small NGO, BUKO, from Hamburg, which had attempted to co-ordinate the very large number of small and disparate

German NGOs in the run up to the Conference, and were running the NGO office, producing and distributing new releases and flyers for meetings. At the top level was the International NGO Steering Group. A crisis at one level would require intervention from the level above, usually itself overstretched.

Thus the overall leadership of the movement networks in Leipzig fell to the International NGO Steering Group, consisting of a group of four people - Henck Hobbelink, Pat Mooney, Vandana Shiva and Liz Hosken of Gaia. The GRAIN-RAFI axis represented the oldest most recognized counter-expertise in the entire field. However, they needed a wider sweep of influence to take the South into the global reach of the anti-GM movement, especially given the problems they had faced with Latin American NGOs at the earlier preparatory meeting in Rome. Vandana Shiva and Tewolde Egziabher (who moved easily between the NGO world and the G77 countries) had played a major mediating role. Thus it was not surprising that Vandana Shiva played a central role, both as a charismatic Southern intellectual and as a boundary spanner through her position in the Third World Network, with its presence in India and SE Asia. Unlike GRAIN and RAFI, who had particular interests and stood for particular positions, Gaia was an NGO committed to networking, particularly with Southern groups. Gaia had strong links in Latin America and with Shiva and the Third World Network. Thus Liz Hosken (Gaia) played a key role as broker between the two networks, North and South. She also acted as a facilitator, bringing together the other three leaders, and at times recruiting volunteers for press work in the NGO office, when the system appeared to be failing.

Immediately following the end of the *In Safe Hands* conference, the NGOs held their first strategy meeting. Two key shifts occurred in this meeting. The NGOs moved from consensus building to preparing for active engagement with their adversaries, and the authority of those leading the movement was consolidated. The NGOs acknowledged the necessity of the leaders using their skills and political experience in the lobbying phase. Thus the meeting started quite casually with Pat Mooney and Henck Hobbelink sitting together at one end of a large oval table, throwing out possibilities for the meeting to discuss. They allowed the NGOs to choose lobbying tactics as the main priority for discussion. An electric atmosphere of anticipation built up as the meeting moved towards the deadline for leaving the building, as more and more people arrived, filling the standing room, as well as the table. Pat Mooney outlined what was needed including five topic areas for lobbying that needed co-ordination. They left it open to the floor to come up with people to fill these posts.

At this point came a key intervention from Daniel Querol, a Nicaraguan NGO, who had been doing the Spanish translation '*You tell us; you know the people*'. He crystallized the trust placed in Pat Mooney and Henck Hobbelink to provide effective leadership to the movement. The force with which he spoke as a Latin American symbolized an effective global consensus around the GRAIN-RAFI axis. For he linked the North to the South, but particularly to Latin America, so recently on its own separate track. His linguistic skills in English (and German), as well as Spanish, allowed him to form a bridge, delivering a degree of coherence to the movement. Given the go-ahead, Henck and Pat allocated responsibilities. Interestingly, they remained very relaxed throughout a meeting which was developing a strong political charge, giving the impression that they knew what they were going to do, whether they carried everyone along with them, or not.

Action Repertoire: Engaging Adversaries

The NGOs occupied a position of marginalized inclusion within the organization of the FAO Conference, symbolised by a whole series of qualified rights. They could enter the plenary, but not the working groups. They could enter the conference hall, but had to sit at the back. They had an office, but inadequate photocopying and no e-mail. They were allowed to organize meetings, but had to hold them in a noisy, smelly tent in the car park.

However, the NGOs were far from insignificant, they maintained a constant presence, a global civil society encampment, keeping a fortress of state power under siege. In prophetic style the NGOs were determined to reverse the centre-periphery relationship, to claim the centre for themselves and treat the intergovernmental business as a bureaucratic periphery. Hence they took a tactical decision to spend the first two days making Farmers' Rights the issue of the conference, while delegates were tied up with the protocol of opening statements. This would then give them command of the ideological heights, from which to criticise the delegates for their lack of progress on resolving the 'real issues'. The NGOs organized a string of fringe meetings, briefing statements and press releases every day to establish a focus on their own issues, to try and set the agenda at the level of ideas rather than procedures. They continuously handed out leaflets and stickers to the delegates in the coffee bars and the atrium outside the conference chamber. Visible lobbying was more restricted to specific events. Attendance at the meetings and press conferences were variable, but usually included NGOs, press, members of International Organizations, such as UPOV, and some governmental delegates, though these were usually from Southern countries, or on occasions the Nordic countries. Leaflets were usually well accepted, but there is no guarantee that they were read. Certainly, NGO advocacy of the importance of the role of women in in-situ conservation of biodiversity, could not have had much resonance with the vast majority of delegates, who conformed to the stereotype of suited, middle aged, male bureaucrats.

On the Friday morning, delegates were taken to an agricultural fair, referred to as the 'Field Days', where, according to the NGOs, they were going to be exposed to a hard sell for industrialized agriculture. The NGOs decided to picket. This provided the NGOs with an opportunity on the one hand to confront the delegates *en masse* and secondly to bond in more disruptive collective action. It was clear form the outset that the most significant players on both sides were absent. The American delegation stuck to their work and the GRAIN-RAFI-Gaia axis (Pat Mooney, Henck Hobbelink, Liz Hosken, Michael Flitner) remained behind to monitor and lobby. Thus, of the leadership, only Vandana Shiva issued forth to rally the troops.

NGOs and students, wearing sandwich boards and hats, formed a human funnel through which the delegates had to pass, in order to enter the show grounds. The demonstrators chanted slogans in English and Spanish and thrust leaflets in their hands. The moment of contact was brief, and all of these delegates (often the more sympathetic Southern ones) would already have been leafleted, invited to fringe meetings and possibly lobbied during the week. However, the tenor of the meeting was very different to that of one-to-one lobbying. The anti-GM movement appeared as a unified group viz a viz the delegates. As the delegates disappeared into the show ground, a Bolivian folk band appeared and the protestors began to dance, led by the Latin Americans and watched by bemused German farmers. After some time of this revelry, the Field Days organizers handed out applications for free tickets. The NGOs held a quick meeting and decided to refuse the invitation. This action presented several opportunities to the anti-GM movement actors, first confrontation, then revelry and finally refusal of an offer. It had an immediate effect of unifying the NGOs and giving them a sense of purpose at a point when the conference appeared deadlocked. Once again this event demonstrated the duality of the counter-expert identity. With the common language of expertise threatening to break down, the activists returned from their temporary status as insiders to the more familiar territory of being outsiders. This action allowed NGOs to embrace their marginality[23], and also through the use of music and dance to move beyond a claim for recognition to an autonomous definition of their identity, and their links with the peasant struggles for survival.

The most dramatic encounter between the prophetic movement and its priestly adversaries, however, occurred in the conference chamber itself. On the final day of the meeting, the NGOs presented a three part attack. Vandana Shiva opened on behalf of the non-profit NGOs, then Pat Mooney closed the focus down to the USA and Canada. Finally Kristin Dawkins, on behalf of the US NGOs, directly attacked the USA for holding up the whole process. The

---

[23] A-SEED students were significant contributors to this action.

NGOs claimed the moral as well as theoretical high ground through their ability to articulate wider ethical issues, of justice and sustainability, to speak freely and forcefully on substantial issues, and against the emptiness of the formal procedures. They demonstrated their courage in confronting power, and speaking plainly on matters of importance. The power of this intervention demonstrated the dimensions of counter-expertise. For not only were they able convey a much stronger grasp of the implications of FAO decisions for the fate of farmers and crop plants than the delegates, but crucially, they were able to appear as 'legitimacy brokers' for the decision making process as a whole.

The NGOs were engaged in defining and leading global civil society, raising conflicts on two fronts. First was their attempt to decouple 'the public interest' from that of the private interests of companies. The NGOs laid claim to representing global civil society and equated the global public interest with global civil society. In so doing, they opposed the clustering of private interests of companies as the economic interests of particular countries, such as the USA, Canada or the UK. Global civil society was set up against the hegemonic coalition of companies and states as the legitimate voice of the public interest. The NGOs worked constantly to present the USA as putting the world at risk for selfish gain, in order to de-legitimate the USA and therefore to block any hegemonic identification between the interests of the US companies and global civil society. Kristin Dawkins attempted to peel a further layer of legitimacy from the USA, by publicly dramatising a conflict within the USA between the economic interests of American agro-business, supported by the US government versus civil society consisting of the American public, farmers and NGOs. At stake was the legacy of Kantian ethics, in which legitimacy depends upon the ability of an actor to reach beyond their own limited interests, and present disinterested, potentially universalizable ethical claims.

Second, the NGOs' prophetic style of leadership clashed with the delegates' priestly style. The former articulated substantive ethical claims, while the latter's diplomatic procedurialism was mainly concerned with the preservation of the organization and the offsetting of personal responsibility (Douglas and Wildavsky, 1982; Douglas, 1993). This encounter dramatised a stark contrast of the charismatic authority of a social movement against the legal-rational authority of a system, as sources of legitimacy (Alberoni in Diani and Melucci, 1988). Charisma allows the leader to act without constraints of tradition or instrumental rationality, the competing forms of legitimacy. Charismatic leadership creates a deep affective resonance that inspires followers to ignore the routines of everyday life and the calculations of costs and benefits. Charisma creates a sense of the exceptional, the revolutionary (Moscovici, 1993) the carnivalesque (Bahktin, 1984; Scott, 1990). The charisma of the three NGO speakers did not depend on irrational identification, but on their embodiment of broadly accepted virtues, which resonate in a way that creates a public. This intervention highlights the prophetic role of social movements (Douglas and Wildavsky, 1982; Douglas, 1993; Melucci, 1996), through which movement actors break the limits of a system, making visible the power that resides in technocratic discourse (Melucci, 1989; 1996). Shiva's blunt description of the conference as political, not technical, drew attention to the unavoidable conflicts involved. Simultaneously it signified a more Nietzschean creative autonomy, breaking out of the Hegelian dialectic of desire for recognition indulged in by the delegates (Der Derian, 1992).

The presence and intervention of the NGOs served to disrupt the smooth functioning of the FAO, and challenge the legitimacy of the collaboration between private financial interests and bureaucratic inertia on behalf of the excluded global civil society.

Evaluating Outcomes and Reaffirming Agency

'Farmers' Rights' presented the core definitional problem for the editorial work of the FAO conference. Farmers' Rights was a concept that was put forward by NGOs in 1985, and had been defined in an FAO document (FAO Resolution 5/89) as the right of farmers to replant saved seed. Nonetheless, Farmers' Rights took on the shape of an 'undecidable'. The chief negotiator from the USA said she did not understand what Farmers' Rights meant and hence the US had substituted the phrase 'the concept of farmers' rights' in a key clause. The G77 countries and the NGOs on the other hand were keen to extend the definition of Farmers' Rights given by the FAO to include benefit sharing and rights to sell seed as farmers chose. Furthermore, they claimed that farmers had always had these rights and that the dispute was around the attempt to remove their rights. Thus they proclaimed that 'Farmers Rights' was not a concept but a reality, and resorted to placing any reference to'the concept of' Farmers' Rights in square brackets. At the beginning of the Leipzig meeting the text of para 62 of the draft Global Plan of Action read:

> To encourage the concrete recognition of [the concept of] farmers' rights at the international, regional and national levels (FAO, 1996a, ITCPRG/96/5 para 62).

In the next version the two contending positions were presented alongside each other as alternatives, in conjunction with the FAO's own position:

> To [realize the concept of Farmers' Rights] [take into account Farmers' Rights] as defined in FAO Resolution 5/89 at the international, regional and national levels (FAO, 1996a, ITCPRG/96/5-Rev 2).

In point of fact, the seed saving practices of farmers were not in the past framed as rights, but taken for granted within particular customary life worlds. The reflexive articulation of seed saving within the Enlightenment discourse of human rights had only appeared as a response to the formulation of corporate powers as Intellectual Property Rights or Plant Breeders' Rights. That is, NGOs were not working within any vernacular moral language of traditional farmers, but lodging claims on behalf of farmers for their recognition in the moral master frame of rights discourse. As I have argued in chapter four, the NGOs tried to combine claims to two different types of rights into the concept of Farmers' Rights: rights to the recognition of farmers autonomy and creativity, and rights of farmers to equity in the distribution of benefits arising from genetic resources. Farmers' Rights was therefore a key concept in the anti-GM movement's injustice frame. Nevertheless, however epistemologically dubious the NGO claims were, the wording of para 62 became the focal point of the struggle between a hegemonic force and its opposing social movement. It took twelve hours behind closed doors on the final day of the conference for a final compromise to be negotiated, in which para 62 read, 'To realize Farmers' Rights, as defined in FAO Resolution 5/89' (ITCPRG/96/5/Rev 2 - Add. 3, Para 62). However, farmers' rights were specified more closely under a different paragraph on benefit sharing:

> Confirming the needs and *individual rights of farmers and, collectively, where recognized by national law, to have non-discriminatory* access to germplasm, information, technologies, financial resources and research and marketing systems necessary for them to continue to manage and improve genetic resources (FAO, 1996a, ITCPRG/96/5/Rev 2 - Add. 3, para 18, bullet point 3).

A tension existed in the concept of Farmers' Rights, in that rights usually refer to individual rights. NGOs were trying to defend collective property from individualization, yet collective rights are more problematic to define. The individualizing pressure the USA brought to bear on their concept of Farmers' Rights was potentially devastating to a project of collective rights for collective property. In the aftermath of the final agreement, a German NGO organizer wore a badge bearing the legend 'Farmers Rights is Lost'. However, Tewolde Egziabher, the lead negotiator for Ethiopia was upbeat. He argued that the form of words (and the lower case) used allowed each country to apply the recognition of Farmers' Rights in the way appropriate to its land tenure system, and not have to stick to a universal form tied to earlier FAO resolutions. Although NGO opinions differed somewhat over the final outcome of the FAO's deliberations, a fairly general feeling was that it was a compromise, which in effect left a lot open for subsequent meetings to decide.

If it was difficult to evaluate the ambiguous legal outcome, NGOs were less ambivalent about there own work. At a final meeting they unanimously decided that the anti-GM movement had come out of the week stronger than it went in. Judging outcomes of social movement activity is a notoriously difficult theoretical problem, to which Diani (1996) has suggested that outcomes are best discussed in network terms. The NGOs in fact chose to adopt a similar position. They emphasized the strength of the NGO / civil society presence, including the scale of the mobilization. More NGOs were present than had ever gathered in once place on this issue before. They stressed the vigour with which the NGO had presented their arguments, the fact that the *Safe Hands* meeting had adopted a unified position, as well as the harmonious working together, sharing of skills and trust. Plans for the NGO mobilization at the next meeting, the FAO's World Food Summit in Rome in November 1996, began at this last strategy meeting in Leipzig. Vandana Shiva coined a new name for the network in the restaurant that evening, Agro-Biodiversity Coalition, with the acronym, ABC, standing for a return to basics. The UK NGO group did indeed adopt this name as a sub-group of the UK Food Group in Rome. NGOs judged their time in Leipzig as a success not so much in terms of their impact on the seed regime, but mainly in terms of their promotion of global civil society as a counter balance to the corporate forces and their statist allies.

New Leaders, New Action Repertoire: Beyond Counter-Experts

While the NGOs maintained a consistent presence as a fringe event to the FAO conference, the fringe too had its own outer edge. A younger generation of movement leadership, more connected to direct action, DIY Culture and expressive forms of soical movement were represented by the European hub of A-SEED (Action for Sustainability, Equality, Environment and Development) and Play Fair Europe. Under the title *Brave New World - International Seminar on Biotechnology, Poverty and the Environment,* they organized a parallel set of seminars in the campsite, which were addressed by several of the leading NGOs, including Pat Mooney. The two organizations had overlapping networks, though Play Fair Europe, in theory, had a distinct educational role. The individuals involved were a network of globalized student and post-student activists[24].

---

[24] The core members were five German students in their mid-twenties (three male and two female), plus a Spanish (male) student resident in Germany; two Britons (both male) - one working in Amsterdam, the other living in Slovenia, using a cyber-cafe as his connection point. He was preparing to move to Rome, in order to work for A-SEED

The A-SEED action repertoire was far more experimental than that of the established NGOs. In addition to the seminars, which were concentrated in the first part of the week to get them up to speed on the issues, they spent a lot of their time in marathon meetings based on consensus decision-making. The explicit purpose of these meetings was to plan their intervention in the FAO, consisting of a speech and an action that evolved into the unfurling of a banner in the conference hall. Their meetings also had to plan future actions at the World Food Summit in Rome in November 1996 and generally to orient their organization for the following year around a 'campaign against globalization and neo-liberalism.' However, most important of all was the process of sorting out 'problems in the group.'

The problems seemed to coalesce around what was effectively a split between two factions, which combined what were perceived as a number of cleavages. The Anglo-German group perceived themselves as older and more experienced, and saw the predominantly female Latin group as younger, more passionate and prone to rhetoric. Cultural differences between German practicality and Latin discursiveness were also mentioned as a problem. They accused some of the Latin women of objecting to decisions that had been taken at meetings they had chosen not to attend. The Latin group argued that they could not be expected to accept decisions if they did not feel they owned the decisions because they had not been properly part of the decision making process.

One such decision was over the speech that A-SEED were to make to the FAO plenary meeting. After an initial five hour meeting of the A-SEED group, they delegated three people to write the speech from the points they had agreed, then the sub-group left one person (a British journalist) to write it, which he did in English. A second meeting the following day, lasting another six hours, had to agree the text and decide who would give the speech in what language. During this time several possibilities were considered, such as two people speaking in different languages, or two in the same language. These

---

along with the Mexican woman, organizing their 'Hunger Gathering' outside the FAO Food Summit.; A Spanish (female) lawyer from Madrid, looking to work in Latin America; A Catalan (female), who described herself as an artist and a Mexican (Female) both living in Amsterdam; and a New Zealander (female) currently living in Prague and editing the Czech Journal of Sociology. On the periphery were myself, a French PhD student (female), who was using A-SEED as a cheap way to attend the conference and a middle aged Dutch 'economist' who claimed to have a novel solution to the environmental effects of current economic models. This last problematic and emotionally demanding figure caused some difficulty in a group with such intense group processes.

options were rejected, only to be reintroduced and rejected again. Finally, the group decided that the English text should be translated into Spanish by a native Spanish speaker, but the speech should be read out by a German woman, who spoke English as her second language and Spanish as her third. The rationalization for this decision was that it would indicate diversity, since neither English nor Spanish are universal languages. This constituted a victory for the sole Catalan woman in A-SEED, who spoke Catalan as her first language, Spanish as a her second language and English as her third, and succeeded in having her own ethnicity dramatised through the form the speech eventually took.

The substance of the speech was that as young people with a stake in the future (a group otherwise completely marginalized from the FAO and global governance) they were critical of the irresponsibility of the older generation who were unable to come to an agreement. The presentation of the speech to the FAO faced technical hitches exacerbated by A-SEED's cumbersome decision-making process. The conference chairperson decided to delay the time of their contribution until after midnight on the final evening. By then, the meeting had, in fact, reached an agreement, reducing the impact of the speech. Though fluent in Spanish and English, the speechmaker was unable to modify the text to suit the changed circumstances. Furthermore, most delegates were not listening to the 'original' Spanish version, but to the English version via the translation service. Since A-SEED had neglected to supply their English version to the translation service, the poor simultaneous translation acted as an extra filter, between A-SEED and its audience.

The rationale for this cumbersome decision-making process was that it allowed everyone to speak and establish a true consensus that every member could own. However, in observing the closing stages, my impression was that the final decision was fairly arbitrary, as exhaustion caused wild swings producing not a gradual zeroing in on the underlying consensus, but a chaotic oscillation, with some members burning out. At several points a decision appeared to have been taken, followed by some people quickly changing position and throwing the meeting back into flux. This fear of closure indicated that the mammoth meetings provided at least some of the A-SEEDers with pleasure and the chance to explore their feelings for each other, their sense of group solidarity, or aspects of their own identity. The debate ended due to the exhaustion of the participants rather than a resolution of the issues.

Affectual flows were of course not contained within meetings. The evenings of drinking, dancing and flirting, with changing patterns of sexual allegiances, spilling over into the NGO community, went on most of the nights. Again, what characterized the A-SEED network was an experimental edge, where no one was settled into a fixed identity. At times they appeared inward looking, engaged in endless internal negotiations. While the other NGOs skirmished with the FAO, they were depleted by the work of maintaining their own collective identity. The collective manual work of washing up, moving exhibitions, and making banners and hats for the Field Days demonstration, acted as a unifying point for them. The A-SEEDers embraced their marginality, as a part of their identity, which included a measure of scepticism towards the value of working on the inside.

A-SEED was not, however, an exclusively expressive grouping. They organized direct actions at the FAO World Food Summit in Rome in November 1996 and subsequent WTO meetings in Madrid, as well as launching an extended e-mail discussion to refine the strategy for their longer-term campaign against globalization and neo-liberalism focusing on the WTO. They were a mobile European protest directed at the embodiment of the powerful forces orchestrating the globalization of the economy. A-SEED embodied the emergence of a global civil society in two ways. Firstly, they focus exclusively on global re-structuring and global institutions in their campaigning work. Secondly, they recruited their members from a number of countries in Europe and beyond. Thirdly, they had globalized mobile lifestyles. Many of them were living outside their country of origin and had the prodigious linguistic skills needed to maintain themselves within this cosmopolitan milieu, as well as their commitment to numerous events in various countries. They were taken very seriously by prominent NGOs, who saw them as a kind of NGO youth wing, free to be more outrageous than the NGOs themselves. A-SEED was appreciated as a source of future activists and possibly also as an experimental cutting edge for the movement.

In November 1996, both NGOs and A-SEEDers congregated in Rome for the FAO's World Food Summit. It was there that larger numbers of the direct action activists fresh from the UK anti-roads movement were recruited to the GM Food debate. These activists returned to the UK with a concern about GM food just as many anti-roads activists were looking for a new issue to move onto and fed into the rapid growth and change in the anti-GM movement in the UK.

Legitimation Stripping in Seattle: Global Civil Society and the WTO

Yet the pattern of global civil society protesting outside global meetings really emerged in Seattle in 1999, when the WTO returned to the negotiating table to set in train a whole new Round of trade talks. The protests were split into two streams, the large official trade union demonstration (approximately 50,000), safely contained well away from the WTO conference and the social movement direct action rebellion, involving about 2,000 protestors (*New Left Review*, 1999) focused on blocking access to the conference venue. Protestors stopped the opening and closing ceremonies of the WTO Ministerial meeting. Most significantly they clearly had the effect of empowering delegates from developing countries (particularly the Africa Group) who refused to accede to the usual bullying of the USA delegation. The WTO had, for the first time, become a public forum where global civil society was gaining access. The counter-expert NGOs formed the thin edge of the wedge, followed by wider protest constituencies, which drew in the global media and therefore a global public. The WTO's customary decision-making methods of backroom deals between the powerful states in regular consultation with the richest companies were suddenly stripped of legitimacy by the public gaze, intensified by globally televized pictures of US police beating and tear gassing protestors in a vain attempt to discipline civil society.

The American press likened the Seattle confrontation to the Battle Chicago in 1968, at the birth of the new social movements. Significant differences were noted. Chicago focused on Vietnam, whereas the Seattle protest drew in many issues. In Chicago the unions, which had sided with the police and the Vietnam War, whereas in Seattle the unions were with the new social movements against the WTO (Gitlin, 1999). The juxtaposition of steelworkers concerned with threats to jobs in the USA with Green protestors in dressed up in Turtle suits was a favoured image of Seattle. The broader constituencies and more complex issues represented in the mobilization in Seattle led to fragile alliances based on potentially conflicting frames and mutually exclusive networks, linked perhaps only by their common adversary. Even leaving the large union presence aside, there was quite a diversity of actors engaged in the direct action protest.

> Who were these direct action warriors on the front line? Earth First!, the Alliance for Sustainable Jobs and the Environment (the new enviro-steelworker alliance), the Ruckus Society (a direct action training center), Food Not Bombs, Global Exchange and a small contingent of anarchists dressed in black, with black masks, plus a hefty international contingent including French farmers, Korean greens, Canadian wheat growers and *British campaigners against genetically modified food* (*New Left Review*, 1999, emphasis added).

The action repertoire of the protests could be separated into four streams. The union march that demonstrated the logic of numbers (della Porta and Diani, 1999) of 'active citizens' from the mainstream of civil society (Moyer, 1990), but was not disruptive. The non-violent occupation of streets and blockading of the conference site by direct action 'rebels' (Moyer, 1990) was designed to disrupt the process of the conference. Spanning the boundary between these two camps was the iconic praxis of 'Mardi Gras' - fancy dress sea turtles and half naked protestors with slogans painted on their bodies who used wit and irony to attract media attention to their imagery. The disruptive action or legitimation stripping (Welsh, 1988) focused media attention on the bodies of the protestors as subjects of discipline and punishment. The police found themselves unable to rely on internalized social discipline to control the protestors and thus moved to physical punishment with tear gas, batons and rubber bullets (*New Left Review*, 2000). The use of violence by the police towards protestors attracted the media, which translated the process of legitimation stripping back from physical violence to symbolic contest. A wider public was exposed to the spectacle of political authorities acting as if they lacked legitimate consent from civil society and were able only to continue in power through direct force. The WTO's loss of credibility, due to over-identification with an American agenda, was compounded by the internal dispute between US government and American civil society on the streets. Institutions of global governance derive their legitimacy from national governments roots in civil society. In this case the direct action separated the national and global power of states from support in global civil society.

The final thread of the action repertoire in Seattle was more confrontational (e.g. looting Starbucks). As always opinion divided between those who thought it represented a radical and fairly restrained response to police provocation and corporate greed (*New Left Review*, 1999) and those for whom such 'negative rebellion' (Moyer, 1990) undermines the movement's base in civil society.

> The masked ones are the outriders of social movements. They can't organize large demonstrations themselves ... their tactics predictably drive the police and the media wild, whereupon the 99 percent of the demonstrators who are non-violent lose control of the event (Gitlin, 1999: 2).

While media may frame protestors who are clubbed and tear-gassed as innocent victims, protestors who take any precautions such as wearing gas masks or helmets are deemed to be contributing to the violence. Nevertheless the global institutions that oversee trade, environmental and other regimes are relatively unaccustomed to dealing with global civil society. As the anti-GM movement's action repertoire expands from counter-experts lobbying, into various types of bigger mobilizations, global civil society is becoming a firmer reality.

Summary

Global governance is a field on which hegemonic projects are played out. There are numerous global institutions. Their connections and ordering are organized by the play of power through the whole field of global governance. An aspect of this hegemonic play is the inclusion of diverse national interests and political viewpoints, balanced against the institutional concentration of power is certain centres and the marginalization of others. These hierarchical power relations are written in the organization of international conferences, such as the FAO conference on Leipzig, and more obviously in Seattle with the WTO. The NGOs of the anti-GM movement inevitably were situated as marginally included in the conference. The national delegates and NGOs offered two culturally different models of global leadership on the issues of biodiversity and biotechnology - what Weber called the priests versus the prophets. The NGOs attended to the usual processes of social movement building maintaining and connecting networks, aligning frames and employing a counter-expert repertoire of means of communicating with their adversaries; and finally evaluating their outcomes and reaffiming the agency of the movement in preparation for further events. The NGO gathering also provided space for a new generation of movement leaders to emerge who were to connect the first generation leaders of the anti-GM movement with new constituencies of direct action protestors. This was to lead to an expansion of the action repertoire of the anti-GM movement both on a national basis in the UK and as part of the global civil society that was to become much more apparent in Seattle in 1999. The anti-GM movement was a pioneer in the creation of this global civil society.

7 Conclusion: Counter-Experts, Social Movements and Global Politics

7 Conclusion: Counter-Experts, Social Movements and Global Politics

The material gathered in this case study of the emergence of the anti-GM movement allows me to address a number of theoretical issues that were raised in chapter one, concerning expert systems and counter-experts; social movements; and global governance, including regimes, global civil society and hegemony.

Expert Systems and Counter-Experts

Dominant theories of contemporary modernity (Giddens, 1990; Beck, 1992; Beck, Giddens and Lash, 1994) have stressed the sequestration of control over everyday life by experts. The individual lay person is confronted by expert systems, which present at best a choice between experts. The significance of counter-experts breaks up a theoretical dependency on a simple expert-lay divide. Instead a much more complex picture emerged from my case study, in which knowledge was negotiated, pieced together from different sources, and inflected with varying political significance. While falling outside the expert realm, the counter-experts were not isolated. On the contrary they had established networks to accommodate the flows of information between them, and to conserve their resources of knowledge, as well as genes.

The counter-expert NGOs who formed the leadership of the anti-GM movement in the mid-1990s could conveniently be split into large membership organizations or small knowledge intensive groups with little or no direct membership. The large NGOs were fairly marginal to the leadership networks, and where they were present, they depended on one or two individual leaders. So the networks were largely personal and depended on the trust each individual placed in others who had proven they could deliver the goods. Many of the key players had been influenced or recruited by a single individual. NGOs combined a level of expertise, with an alignment in an oppositional movement, which led me to call them counter-experts. Three analytical dimensions of counter-expertise are significant for my account: cognitive, political and epistemological.

- *Cognitive*. Counter-experts have to have a high level of knowledge or skill in the field. They must be conversant with the science and the law, so that they are able to engage in expert debates. To be taken seriously they must often prove themselves as knowing more than the experts do.

- *Political*. The second dimension requires that they are aware of the socio-political context and the implications of technologies. In tracing the complex implications left out of reckoning by the industry's experts the counter-experts are attuned to the unease of diverse publics confronted in different ways by new technologies. This awareness, combined with a structural position outside the expert systems of industry and government, gives the counter-experts the legitimacy of representatives of the public interest. Counter-experts are thus able to act as legitimacy brokers in international organizations, which have little direct contact with civil society.

- *Epistemological.* While expert knowledge claims are in theory uncertain (Giddens, 1990) they appear in the wider public domain as certain bases for wide-ranging technological developments and policies. Counter-experts confront technical claims with higher levels of uncertainty of outcomes and feed wider social unease about the new technologies and IPRs. Critique of positivist science and policy-making is a fundamental part of the counter-expert role. The NGOs' ability to live with uncertainty is linked to their effectiveness within chaotic global systems. NGOs are able to achieve a 'butterfly effect', combining skill, legitimacy and awareness of uncertainty to intervene in global politics out of all proportion to their small size. As the NGOs become increasingly expert, it becomes easier to communicate with adversaries than with supporters.

These counter-experts became the leadership around which a new protest cycle has developed. Yet their ability to frame issues and translate between wider public concerns and more institutional discourses will remain important, even though other forms of leadership may be growing in importance as well.

Social Movements: Networks, Frames and Repertoires

Social movements are continually changing. There is no mileage to be gained from applying analytical categories of progressive versus defensive (Habermas, 1981; Offe, 1985) or using the 1970s as a measuring stick (Jamison and

Eyerman, 1991; Jamison, 1994; 1996) for movements in the 1990s and after. It has been argued that the concept 'social movement' is primarily descriptive rather than defined by a single analytical distinction (Yearley, 1994). Even 'the environmental movement' is impossible to define in the singular. The anti-GM movement is one of numerous movements that are environmental. Environmental movements overlap with 'third world development' and feminism, but also with the New Age movement, not conventionally considered a new social movement (Heelas, 1996). Collective action will not necessarily occur in familiar forms (Touraine, 1992). The 1990s witnessed new modular forms of site-specific direct action, as well as new interventions in everyday life and global negotiation. The anti-GM movement has itself changed its form with new waves of mobilization.

The inequalities of power and sovereignty represented in global institutions such as the WTO means that a social movement on an issue such as seed patenting cannot be constituted on a national basis in a Southern country alone. Its adversaries are global in reach but primarily located in the North. Social movement activity has not remained as experimentation on the local margins, but has gone to the centre of global institutions, where new forms of global governance and hegemony are being shaped. The activists I have described have shown the way towards a new political terrain and new types of social movement.

- They renewed global institutions by introducing the concerns of global civil society into these previously exclusionary interstate institutions.

- They renewed Northern environmental movements by opening them to Southern influences.

- They re-framed everyday life practices in North and South in terms of global issues of biodiversity conservation and social justice.

The common ground between different social movement theories has been summarized as suggesting that social movements (especially the more political ones) possess four common characteristics. '(1) Informal networks, based (2) on shared beliefs and solidarity, which mobilise about (3) conflictual issues, through (4) the frequent use of various forms of protest' (Della Porta and Diani, 1999: 16). From a specifically New Social Movements perspective Melucci suggests that social movements express solidarity; articulate a conflict; and break the limit of the system (Melucci 1989; 1996). These criteria direct attention to networks that embody solidarity; frames that shape value judgements about a conflict; and action repertoires, respectively.

Solidarity Networks

Social movement networks depend on flows of information and affect between bodies. The NGO counter-experts are dispersed across the globe and keep in contact in ways that are familiar to academic and professional networks: publishing, phoning, faxing or e-mailing each other, with face-to-face contact at conferences and meetings. Flows of information are often channelled through a coalition around a specific issue organized by an individual with a particular interest. A crisis, or sudden change, such as occurred with the European patenting directive, can create a cascade of rapidly widening activity as the networks are activated.

Large NGO meetings or major international conferences give NGOs a series of ways of developing trust and solidarity through social contact around the business itself. Eating, drinking, dancing and talking create a collective effervescence, allowing flows of affect to bind the seed networks together. The chance to meet, work and socialize together consolidates the solidarity between seed activists and their shared ways of framing seed issues.

Framing Conflict and Value Judgements

As movement leaders, one of the great strengths of counter-experts is developing discursive frames. The NGOs produced a 'Green Radical discourse' (Dryzek, 1997) on biodiversity and biotechnology. The NGOs worked hard at defining the conflicts they were engage in, which required the recognition of adversaries and issues at stake in mapping out the contested social field. Considerable work at conferences, meetings and in publications was aimed at linking issues and to some extent in separating them, so that biodiversity could be shifted from mega-fauna to food crops; the natural to the social. Adversaries were identified as a coalition of actors engaged in promoting and patenting GMOs. Transnational agro-chemical corporations lie at the heart of this coalition, but connect to national governments such as the USA and Canada, international organizations such as the WTO, and the whole expert system of biotechnology. Each of these can become the prime focus at a particular time for a particular NGO, as can the EU seed quality regulations (Purdue, 2000) or an advocate of biotechnology such as the Science Museum (Purdue, 1996; 1999).

The critical discourse, in which the NGOs framed the concerns of the anti-GM movement, was built up from a combination of a 'risk frame' and a 'rights frame'. The risk frame poses biotechnology as a direct risk to human health in the form of medical biotechnology and through the potential health dangers of GM food. Environmental risks covers a wider range including the impact of GMOs on wild species. More pressing is the risk that patented GM crops will drive out other varieties and lead to rapid further decline in agricultural biodiversity. These environmental risks have a potential to effect human health. However, the route from environment to human health goes mostly via the social issues grouped under the rights frame. The concept of Farmers' Rights as a counter-balance to Intellectual Property Rights indicates that patented GM crops limit the autonomy of farmers to pursue their livelihood. It also claims that seed patenting shifts the distribution of economic and ecological resources against equity for farmers, crucially poor farmers in the South, and in favour of rich transnational corporations in the North. The rights frame therefore includes rights to social justice as well as rights to autonomous action.

Action Repertoires: Stretching the Limits

Social movements engage in both participation and protest, depending on contingent factors (Diani, 1997). Action repertoires cannot be simply categorised into conventional versus disruptive. Counter-experts add a new module to the 'modular' action repertoire of collective action, which social movements have been developing since the French Revolution (Tarrow, 1994). Set in a global context, counter-experts represent another important new tendency in contemporary movements, to respond to and even create political opportunities at the global level that more conventional political actors, such as political parties, are unable to exploit.

The NGOs did not break the international state system, in the sense of employing a radically disruptive action repertoire, although anti-GM protestors contributed to the direct action at the Seattle meeting of the WTO. Neither did the NGOs entirely fit the rules of the international game. The very presence of non-state actors in inter-governmental conferences, such as the FAO meeting in Leipzig, undermined the monopoly of representation held by states. The NGOs have stretched and re-shaped the international state system. Neither absorbed nor entirely excluded from global institutions, they occupied an interesting point on a spectrum of marginalized inclusion. Balancing on the edge of global institutions, they functioned as boundary objects (Gieryn, 1995; Washburne, 1997) or more properly boundary subjects, opening the system to scrutiny in which the working of hegemony through international system becomes visible (Melucci, 1988). The degree of inclusion varied from one institution to another, and this in turn, roughly speaking, inversely correlated to the degree of inclusion or marginalization of the institution itself in the dynamics of global hegemony. Thus NGOs were relatively powerful in the Plant Genetic

Resources Commission of the FAO, but less so in the World Food Summit, or the WTO, yet social movement protest was effective at the WTO meeting in Seattle.

Counter-expertise is not only an action repertoire module it is also a leadership practice. Counter-experts formed the anti-GM movement and found a way to communicate with their adversaries. As the movement has developed it has moved into a new phase in which a broader mobilzation has occurred adding legitimation stripping (Welsh, 1988) to the movement's action repertoire and new forms of leadership. Seed savers (Purdue, 2000) continue to build their own cultural enclaves on the allotments, while local GenetiX groups are merging related environmental concerns with the new anti-GM movement.

Global Governance

Global Regimes

Three models of regime formation have emerged from the neo-realist school of International Relations, all of which are state centric and all but exclude the influence of social movements.

- Rational choice: sovereign states act as rational utility maximizers and are committed to regimes exactly in as far as their interests are served, as if the regime were a market.

- Hegemony: a single undisputed hegemonic state has the power to guarantee the terms of the regime.

- Institutional bargaining: states enter open-ended negotiations with some commitment to the norms and values underlying the regime, under a less defined form of entrepreneurial leadership (Young, 1989). In this version a regime resembles a partnership built on trust.

NGO counter-experts play a more significant part in seed regimes than any of these models allow. The classically realist rational choice model is flawed, as states are not necessarily unified and isolated actors, nor do they have perfect information on which to vote. So the NGOs played an important role in supplying information and interpreting legal texts; this allowed delegates to clarify their perception of their own interests. NGOs were also prone to building alliances between countries that they perceive as having common interests or to contribute to the strengthening of unified positions within existing alliances, like the G77 block.

The hegemony model assumes an identity between domination and legitimacy. Hegemony is used to mean the power of the dominant state without the need for legitimation. In all the seed patenting fora, a distance lay between the dominant power (USA) and the views of many of the other states. Hence the USA succeeded in the GATT, but only after applying bilateral pressure outside the trade talks on 25 countries to adopt strict patent laws. When the Biodiversity Convention was first agreed the USA refused to sign. They only ratified the Treaty after a change of President and a rather contentious intervention by three large US NGOs. Counter-expert NGOs acted to align power and legitimacy in the case of the Biodiversity Convention. In the FAO, where US NGOs publicly aligned themselves with the G77 countries, against the government of the USA, they acted to disengage legitimacy from the power of the USA as global leader.

The institutional bargaining model is the most pluralist of the three and can most easily accommodate NGOs or social movements. However, it ignores the fact that a hegemonic power, such as the USA may frequently operate outside a multi-lateral regime, such as the GATT/WTO using considerable bilateral pressure to achieve its policy aims, as was the case with the introduction of the TRIPs agreement. Within the bargaining model, NGOs provided creative leadership on specific issues where states fail to do so or provide a unifying perspective that transcends the limits of national strategies. In the terms I have used, NGOs can provide both expertise and legitimacy to a regime; this requires an open-ended commitment from nation states. At the FAO: NGOs were co-opted as experts to prepare the documentation for the meeting, such as the Global Plan of Action; and NGOs have fed theoretical ideas into the meetings over a long period (for example, the key concept of Farmers' Rights). They also acted as legitimacy brokers, critical of the delegates tied to national and corporate interests.

The Intellectual Property regime governing seeds was complex, because there was more than one institutional focus. Different institutions evolved in different directions. The FAO Commission on Plant Genetic Resources moved from industrial self-regulation towards an environmental regime, while in the WTO economic interests clearly dominated protection of the environment. It was extremely difficult for NGOs with limited resources to intervene in all these global political sites simultaneously. There was no clear and fixed ordering of the importance of global institutions. Hegemony is mobile, centres of power shift from one institution to another. For example the WIPO and the FAO were both marginalized with the creation of the WTO. The capture of an apparent centre of power may prove to be a hollow victory as key decision making powers shift elsewhere.

Regime theory tends towards political reductionism and concentrates on formal institutions and diplomatic deals. However, the resources for a long-term struggle in and around these institutions were arguably being developed in the construction of a wider global civil society. Regime theory avoids the social and cultural impacts of social movements altogether, which are better approached from the perspective of global civil society.

Global Civil Society

Global civil society is becoming increasing vigorous in spite of the technological and linguistic mediations required to maintain it. As the global becomes a familiar cultural terrain, engagement with global politics is likely to increase and concentrations of skill and innovation outside the formal interstate system is likely to become a more accepted part of politics. Social movement theorists will have to remain alert to these rapidly changing social and technological forms if they are to keep abreast of contemporary social movements. Yet, much of the mainstream of social movement theory retains a primary focus on the nation state (Tarrow, 1994). Those of a more globalist bent (Touraine, 1995; Melucci, 1985; 1989; 1996; Hegedus, 1989) are content to emphasize individual ethical responses to self-evidently global problems (Hegedus, 1989) or global risks as a source of planetary consciousness (Melucci, 1985). Little attention has been given to the knowledge that social movement actors have gained about global governance and the skill with which they have intervened. Greenpeace, quicker than most to pick up the salience of global governance, have been heavily criticised for it (Jamison and Eyerman, 1991; Jamison, 1994; 1996). The concept of global civil society is a useful conduit for bringing together global governance and social movements. The counter-expert leadership of the anti-GM movement facilitated the growth of global civil society in three ways.

- The NGO counter-experts formed global networks outside state or inter-state structures. These global networks connected electronically and met face-to-face at movement events allowing them to represent the movement to itself as an autonomous actor in a global social field.

- Counter-experts re-framed everyday habits such as eating and seed saving within a global ecological context. This shifts civil action and associations towards a global context. Protestors, seed savers and consumers act locally, but make their choices as global citizens, with reference to a global frame of biodiversity crisis, risks and rights.

- Finally, action led by the NGOs has established nodes of civil society around state-centric global institutions and negotiations. The NGOs saw themselves as representing new identities and interests in and around the institutions of global governance. Hence their ability to act as 'legitimacy brokers' or alternative conduits for the flow of public opinion, and their recognition as such by state delegates in the international fora.

Global civil society is emerging as a complex web of interactions - some face-to-face, many more mediated electronically. Civil society interacts with more formal politics on the one side and private concerns of citizens on the other. A global anti-GM movement may engage with the negotiating positions of blocs of states or national governments, such as the desire of the Indian government to protect its pharmaceutical industry. However, seed saving is locally nuanced (local seeds for local conditions) in a way that Hegedus's global ethical commitments are not. Closer to the private end of the global civil society spectrum anti-GM NGOs engage with the immediate concerns of local farming communities or the attempts by particular individuals on an allotment in Bath to grow a local lettuce variety. The global is simultaneously local. Seed activists in the NGOs play an important part in the weaving of this global-local web.

A global civil society approach ignores the direct impact that social movements can have on international negotiations, which is better addressed by regime theory. The global civil society approach also ignores the fact that framing issues goes beyond globalising them. Framing involves revalorising key terms, such as whales, rain forests or seeds. Social movements wrap these terms in new clouds of meaning and affect, and develop new identities in which these key terms are a significant stake.

Global Hegemony

Hegemony straddles the divide between formal institutions and civil society where it exists. Hegemony is a structural form of leadership - executive power in the institutions is complemented by popular support in civil society. The extension of genetic patents from their origin in the USA to a global level constituted a global hegemonic project in which the economic interests of the agro-chemical transnationals were framed in an expert discourse of patent law. This hegemonic project was primarily aimed at reshaping institutions at the global level where civil society has historically been very weak. Global civil society emerged as a counterbalance to the monopoly of state power in global governance.

The globalization of this hegemonic project crystallized with the signing of the TRIPs Agreement in 1993, and the formation of the WTO. The TRIPs agreement specified that all the member states had to allow patenting of genes and seeds. It became a reference point for the inclusion of similar intellectual property clauses in other global institutions dealing with governance of seeds, such as the CBD and the FAO. The only loophole in the TRIPs, the *sui generis* clause, has been under pressure to be closed by reference to UPOV. This global enclosure was presented as if it were the logical outcome of the working through of the scientific, legal and diplomatic expert systems. The apparent fusion of the logic of systems with the interests of the dominant actors in those systems was a hegemonic project.

Hegemony at a global level proved to be more complex than formulations in regime theory allowed. Rather than an instrument possessed by the dominant power, hegemony revealed itself to be always an incomplete project, open to contest. Tensions between international organizations allowed different amounts of play. A single regime was extended through several global institutions, with anchor points tying one treaty to another. Precedents for this use of the concept of hegemonic projects have been set in State theory (Poulantzas, 1973) and discourses of national politics (Laclau, 1990; Hall, 1988). Yet despite the distinguished pedigree of the concept of hegemony in International Relations, the concept of hegemonic projects has not been applied in the analysis of global power. Furthermore the concept of hegemony remained, even in post-Marxist hands (Laclau, 1990), wedded to a socialist project of counter-hegemony. In linking hegemony to social movements instead of class analysis, I have attempted to move away from a totalising project and into a territory that lies beyond grand narratives.

In using the concept of hegemony to indicate an always-incomplete project, I have also suggested a solution to the dilemma of what constitutes the adversary against which social movements struggle: whether it is a dominant (collective) actor or the logic of a system. A hegemonic project is a combination of both. Dominant actors assemble hegemonic discourse-coalitions (Hajer, 1995). Biotechnology links clusters of diverse actors - companies, national governments, international organizations, experts, lobbyists - into networks and weave together discursive elements - molecular genetics, patent law, free trade theory - into a discursive frame, and seek support for this position whether they judge is necessary, just as social movement actors do. Is a hegemonic project then a real material object? In some senses yes, but bearing in mind Melucci's reminder that social movements make systemic power visible, it is the recognition of hegemony by its opponents in the social movements that gives it tangible political existence. The anti-GM movement that I have been studying has, with varying degrees of success, challenged the hegemonic project of global patenting both in the negotiation of regimes within international law and by contributing to the development of a global civil society. And so it is the activists themselves that have named the power at the heart of the global age. This book is intended as a contribution to that work.

Bibliography

Acharya, R. (1992) 'Patenting of Biotechnology: GATT and the Erosion of the World's Biodiversity', *Journal of World Trade*, 25 (6): 71-87.

Bahktin, M. (1984) *Rabelais and his World*, Bloomington: Indiana University Press.

Bainbridge, D. (1992) *Intellectual Property*, London: Pitman.

Balibar, E. (1970) 'Basic Concepts of Historical Materialism' in *Reading Capital*, Althusser, L, and Balibar, E, London: NLB.

Bauman, Z. (1990) *Thinking Sociologically*, Oxford: Blackwell.

Bauman, Z. (1993) *Postmodern Ethics*, Oxford: Blackwell.

Baumann, M. (1995) 'First Swiss Plant Variety Patent Revoked', *Seedling*, 12 (2): 14-16.

Baumann, M, Bell, J, Koechlin, F & Pimbert, M. (1996) *The Life Industry:Biodiversity, People and Profits*, London: IT Publications.

Beck, U. (1992) *Risk Society: Towards a New Modernity*, London: Sage.

Beck, U, Giddens, A, and Lash, S. (1994) *Reflexive Modernization*, London: Sage.

Billig (1995) 'Rhetorical Psychology, Ideological Thinking and Imagining Nationhood', in H. Johnston & B. Klandermans (ed.), *Social Movements and Culture*, London: UCL.

Branson, W, and Klevorick, A. (1986), 'Strategic Behaviour and Trade Policy' in P. Krugman (ed.), *Trade Policy and the New Economics*, Cambridge MA and London: MIT Press.

Bryman, A. (1992) *Charisma and Leadership in Organizations*, London: Sage.

Busch, L, Lacy, W, Burkhardt, J & Lacy, L. (1991) *Plants, Power and Profit: Social, Economic and Ethical Consequences of the New Biotechnologies*, Oxford: Basil Blackwell.

CIEL (1999) 'Should the WTO Negotiate New Trade Rules on Genetically Modified Organisms?' Draft Discussion Paper, <http://www.sustain.org/biotech.

Callon, M. (1986) 'Some elements of a sociology of translation: domestication of the scallops and the fishermen of St Brieuc Bay,' in J. Law (ed.), *Power, Action and Belief: A New Sociology of Knowledge*, London: Routledge & Kegan Paul.

Cameron, J & Ward, H. (1992) *The Multilateral Trade Organization: a Legal and Environmental Assessment*, Gland, Switzerland: Worldwide Fund for Nature.

Campbell, B. (1989) 'Generalists, practitioners, and intellectuals: the credibility of experts in English patent law', in R. Smith and B. Wynne (ed.), *Expert Evidence*, London: Routledge.

Chemers, M. (1993) 'An Integrative Theory of Leadership' in M. Chemers and R. Ayman (Ed) *Leadership Theory and Research: Perspectives and Directions*, San Diego and London: Academic Press.

Cherfas, J, Fanton, M, and Fanton, J. (1996) *The Seed Savers' Handbook*, Bristol: Grover Books.

Cheyne, I. (1992) 'Environmental Treaties and the GATT' *Reciel*, 1 (1): 14-20.

Coates, B. (1999) 'Friends fall out', *The Guardian*, 8/12/1999: 4-5.

Constantinou, C. M. (1994) 'Diplomatic Representations ... or Who Framed the Ambassadors?', *Millennium*, 23 (1): 1-23.

Constantinou, C. M. (1995) 'Gastronomic Diplomacy', Paper presented to the School of Politics, UWE, Oct 1995.

Cook, T, Doyle, C & Jabbari, D. (1991), *Pharmaceuticals, Biotechnology and the Law*, New York and London: Macmillan / Stockton.

Della Porta, D, and Diani, M. (1999) *Social Movements: An Introduction*, Oxford: Blackwell.

Der Derian, J. (1992) *Antidiplomacy: Spies, Terror, Speed and War*, Oxford: Blackwell.

Dhanjee, R & Boisson de Chazournes, L. (1990), 'Trade Related Aspects of Intellectual Property Rights (TRIPS): Objectives, Approaches and Basic Principles of the GATT and of Intellectual Property Conventions', *Journal of World Trade*, Vol.24, No.5, pp.5-15.

Diani, M. (1996) 'Social Movement Outcomes: A Network Perspective', Paper presented at the *Second European Conference on Social Movements*, Universidad del Pais Vasco, Vitoria (Vizcaya-Euskadi), 2-5 October 1996.

Diani, M. (1997) 'Organizational Change and Communication Styles in Western European Environmental Organizations', Paper presented to the ECPR Joint Sessions, University of Bern, 27 February - 4 March.

Diani, M & Melucci, A. (1988) 'Searching for Autonomy: The Sociology of Social Movements in Italy', *Social Science Information*, 27 (3): 333-353.

Douglas, M. (1993) 'Governability: A Question of Culture', *Millennium*, 22 (3): 463-482.

Douglas, M, & Wildavsky, A. (1982) *Risk and Culture: An Essay on the Selection of Technological and Environmental Dangers*, University of California Press: Berkeley and Los Angeles.

Dryzek, J. (1997) *The Politics of the Earth: Environmental Discourses*, Oxford: OUP.

Dubey, M. (1992) 'The Final Uruguay Text: A Critical Analysis' *Third World Economics*, Sept/Oct, 1992.

Dyer, H. (1993) 'EcoCultures: Global Culture in the Age of Ecology', *Millennium*, 22 (3): 483-504.

The Ecologist (1990), 'Special Issue on GATT', 20 (6).

The Ecologist (1992) 'Special Issue on the Commons' 22 (4).

The Ecologist (1993), 'Cakes and Caviar? The Dunkel Draft and Third World Agriculture' 23 (6): 219-222.

Eder, K. (1996a) *The Social Construction of Nature: A Sociology of Ecological Enlightenment*, London: Sage.

Eder, K. (1996b) 'The Institutionalization of Environmentalism: Ecological Discourse and the Second Transformation of the Public Sphere', in S. Lash, B. Szerszynski & B. Wynne (ed.), *Risk, Environment and Modernity*, London: Sage.

Egziabher, Tewolde B.G. (1994) Workshop Presentation to *Patents, Genes and Butterflies*, International Symposium, Bern, 21/10/1994.

Ektowitz, H, and Webster, A. (1995) 'Science as Intellectual Property' in S. Jasanoff, G. Markle, J. Petersen & T. Pinch (ed.) *Handbook of Science and Technology Studies,* London: Sage.

Elliot, L. (1999) 'US and EU jostle to steer talks' *The Guardian*, 1/12/1999: 15.

Food and Agriculture Organization. (1996a) *Global Plan of Action for the Conservation of Plant Genetic Resources*, (Final Revision), ITCPRG/96/5/Rev 2.

Food and Agriculture Organization. (1996b) *Leipzig Declaration*, (Final Revision), ITCPGR/96/6/Rev. 2.

Foucault, M. (1977) *Discipline and Punish*, London: Allen Lane.

Foucault, M. (1978) *History of Sexuality Vol I*, London: Penguin.

Fowler, C, and Mooney, P. (1990) *The Threatened Gene: Food, Politics, and the Loss of Genetic Diversity*, Cambridge: Lutterworth Press.

Frisch, M. (1994) *Directory for the Environment: Organizations, Campaigns and Initiatives in the British Isles*, London: Green Print.

Gamson, W. (1995) 'Constructing Social Protest', in H. Johnston & B. Klandermans (ed.), *Social Movements and Culture*, London: UCL.

General Agreement on Trade and Tariffs (1992) 'Draft Agreement on Trade-Related Aspects of Intellectual Property Rights, Including Trade in Counterfeit Goods, Submitted by GATT Director, Arthur Dunkel' *World Intellectual Property Report*, 6 (2): 42-55.

GenEthics News (1997) 'Activists target Monsanto and Sainsbury's', *GenEthics News* 17: 4-5.

Giddens, A. (1984) *The Constitution of Society: Outline of the Theory of Structuration*, Cambridge: Polity.

Giddens, A. (1990) *The Consequences of Modernity*, Cambridge: Polity Press.

Giddens, A. (1998) *The Third Way: The Renewal of Social Democracy*, Cambridge: Polity.

Gieryn, T. (1995) 'Boundaries of Science', in S. Jasanoff, G. Markle, J. Petersen and T. Pinch (ed.), *Handbook of Science and Technology Studies*, London: Sage, pp 393-443.

Gitlin, T. (1999) 'From Chicago to Seattle' *Newsweek*, 13/12/1999: 2.

Goffman, E. (1969) *The Presentation of Self in Everyday Life*, London: Allen Lane.

GRAIN. (1994) 'Towards a World Gene Bank?', *Seedling*, 11 (2): 3-10.

Gramsci, A. (1971), *Selections from the Prison Notebooks of Antonio Gramsci*, Edited and translated by Q. Hoare and G. Nowell Smith, London: Lawrence & Wishart.

Grove White, R, Macnaghten, P, Mayer, S, and Wynne, B. (1997) *Uncertain World: Genetically Modified Organisms, Food and Public Attitudes in Britain*, Lancaster: CSEC.

Guardian. (19/11/1993), 'Christopher Warns of Direct Action Against N Korea'.

Habermas, J. (1981) 'New Social Movements' *Telos*, 49: 33-7.

Hajer, M. (1995) *The Politics of Environmental Discourse: Ecological Modernization and the Policy Process*, Oxford: Clarendon.

Hall, S. (1988), *The Hard Road to Renewal: Thatcherism and the Crisis of the Left*, London: Verso.

Hannerz, U. (1980) *Exploring the City: Inquiries Toward an Urban Anthropology*, New York: Columbia UP.

Hannerz, U. (1990) 'Cosmopolitans and Locals in World Culture', *Theory, Culture and Society*, 7: 237-251.

Haraway, D. (1992) *Primate Visions: Gender, Race and Nature in the World of Modern Science*, London: Verso.

Haraway, D. (1997) *Modest_Witness@Second_Millennium.FemaleMan©_Meets_Onco Mouse™: Feminism and Technoscience*, London: Routledge.

Harris, P. (1994) 'Between Science and Shamanism' in K. Milton (ed.), *Environmentalism: The View from Anthropology*, London: Routledge.

Harvey, D. (1993) *The Condition of Postmodernity: An Enquiry into the Origins of Cultural Change*, Cambridge MA and Oxford: Blackwell.

Hecht, S and Cockburn, A. (1989) *The Fate of the Forest: Developers, Destroyers and Defenders of the Amazon*, London: Verso.

Heelas, P. (1996) *The New Age Movement*, Oxford: Basil Blackwell.

Hegedus, Z. (1989) 'Social Movements and Social Change in Self-Creative Society: New Civil Initiatives in the International Arena', *International Sociology*, 4 (1): 19-36.

Hobbelink, H. (1991) *Biotechnology and the Future of World Agriculture*, London: Zed Books.

Hollander, E. (1993) 'Legitimacy, Power, and Influence: A Perspective on Relational Features of Leadership' in M. Chemers and R. Ayman (Ed) *Leadership Theory and Research: Perspectives and Directions*, San Diego and London: Academic Press.

Inglehart, R. (1981) 'Post-Materialism in an Environment of Security' *American Political Science Review*, 75: 880-900.

Irwin, A. (1995) *Citizen Science: A Study of People, Expertise and Sustainable Development*, London: Routledge.

Jamison, A. (1994) 'Intellectuals and Social Movements: America in the 1950's and Europe in the 1990's', Paper presented at the *Politics of Cultural Change Conference*, Lancaster University, 8-10 July 1994.

Jamison, A. (1996) 'The Shaping of the Global Environmental Agenda: The Role of Non-Governmental Organizations', in S. Lash, B. Szerszynski and B. Wynne (ed.), *Risk, Environment and Modernity*, London: Sage.

Jamison, A, and Eyerman, R. (1991) *Social Movements: A Cognitive Approach*, Cambridge: Polity Press.

Jessop, B. (1995) 'The regulation approach, governance and post-Fordism: alternative perspectives on economic and political change?' *Economy and Society*, 24 (3): 307-333.

Jones, N, (1992) 'Biotechnological Patents in Europe - Update on the Draft Directive', *European Intellectual Property Review*, 12: 455-7.

King, D. (1992) 'Legal, Socio-Economic and Ethical Issues in the Patenting of Plants and Animals', Unpublished Seminar Paper, London: CIIR.

King, D. (1996) 'Industry Watch - the seed industry', *GenEthics News*13, Sept/Oct.

Kloppenburg, J. (1988a) *First the Seed: The Political Economy of Plant Biotechnology 1492 - 2000*, Cambridge: C.U.P.

Kloppenburg, J (1988b) *Seeds and Sovereignty: the Use and Control of Plant Genetic Resources*, Durham & London: Duke University Press.

Kloppenburg, J. (1994) 'Changes in the genetic supply industry' Paper presented to International Symposium *Patents, Genes, Butterflies*, Bern: WWF and Swissaid, 20-21 October 1994.

Ko, Y. (1992), 'An Economic Analysis of Biotechnology Patent Protection', *Yale Law Journal*, 102 (3): 777-804.

Kolleck, R. (1994) 'The Gene: A Diffuse Object of Desire', Paper presented to International Symposium *Patents, Genes, Butterflies*, Worldwide Fund for Nature and Swissaid, Bern, 20-21 October 1994.

Kolleck, R. (1995) 'Ambiguous Genes', *Biotechnology and Development Monitor*, June, p 24.

Korten, D. (1992) 'Development Heresy and the Ecological Revolution', *Development*, 2: 97-102.

Kriesi, H, Koopmans, R, Dyvendak, J and Guigni, M. (1995) *New Social Movements in Western Europe: A Comparative Analysis*, London: UCL Press.

Krugman, P. (ed.), (1986), *Strategic Trade Policy and the New Economics*, Cambridge MA & London: MIT Press.

Laclau, E. (1990) *New Reflections on the Revolution of Our Time*, London: Verso.

Lash, S and Urry, J. (1994) *Economies of Signs and Space*, London: Sage.

Latour, B. (1987) *Science as Social Action: How to follow scientists and engineers through society*, Milton Keynes: O.U.P.

Latour, B. (1988) *The Pasteurization of France*, Cambridge MA: Harvard U.P.

Latour, B. (1993) *We Have Never Been Modern*. Hemel Hempstead: Harvester Wheatsheaf.

Levidow, L (1991) 'Cleaning up on the farm', *Science as Culture*, 2 (4), no 13: 538-568.

Levidow, L. (1996) 'Simulating mother nature, industrialising agriculture', in G. Robertson, M. Mash, L. Tickner, J. Bird, B. Curtis and T. Putnam (ed.), *FutureNatural: Nature, science, culture*, London & New York: Routledge.

Lyotard, J-F. (1986) *The Postmodern Condition*, trans. R. Durand, Manchester: M.U.P.

Lyotard, J-F. (1988) *The Differend*, Manchester: M.U.P.

McDougall, C. (1995) *Intellectual Property Rights and the Biodiversity Convention: The Impact of GATT*, London: Friends of the Earth.

Maffesoli, M. (1996) *The Time of the Tribes: The Decline of Individualism in Mass Society*, London: Sage.

Marx, K. (1976) 'So-Called Primitive Accumulation', *Capital Volume One*, London: Penguin and New Left Review.

Melucci, A. (1985) 'The Symbolic Challenge of Contemporary Movements', *Social Research*, 52 (4).

Melucci, A. (1989) *Nomads of the Present: Social Movements and Individual Needs in Contemporary Society*, London: Hutchinson Radius.

Melucci, A. (1992) 'Frontier Land: Collective Action between Actors and Systems' in M. Diani and R. Eyerman (ed.), *Studying Collective Action*, London: Sage, pp 238-258.

Melucci, A. (1995) 'The Process of Collective Identity', in H. Johnston and B. Klandermans (ed.), *Social Movements and Culture*, London: UCL.

Melucci, A. (1996) *Challenging the Codes: Collective Action in the Information Age*, Cambridge: C.U.P.

Mooney, P. (1979) *Seeds of the Earth: A Private or Public Resource?*, Ottawa & London: Inter Pares & ICDA.

Mooney, P. (1993) 'Genetic Resources in the International Commons' (Unpublished Draft).

Mooney, P. (1994) 'Trade Rules, IPRs and the Biodiversity Convention' Paper presented to International Symposium *Patents, Genes, Butterflies*, Bern, WWF & Swissaid, 20-21 October 1994.

Moscovici, S. (1993) *The Invention of Society: Psychological Explanations for Social Phenomena*, Cambridge: Polity.

Moscovici, S. (1994) 'Three Concepts: Minority, Conflict and Behavioural Style' in S. Moscovici, A. Mucchia-Faina and A. Maas (ed.) *Minority Influence*, Chicago: Nelson-Hall.

Moyer, B. (1987) 'The Movement Action Plan: A Strategic Framework Describing the Eight Stages of Successful Social Movements', San Francisco: Social Movement Empowerment Project.

Moyer, B. (1990) 'The Practical Strategist: Movement Action Plan (MAP) strategic theories for Evaluating, Planning and Conducting Social Movements', San Francisco: Social Movement Empowerment Project.

Myers, G. (1995) 'From Discovery to Invention: The Writing and Rewriting of Two Patents', *Social Studies of Science*, 25: 57-105.

New Left Review. (1999) 'Seattle Diary: It's a Gas, Gas Gas', *New Left Review*, 238: 81-96.

Newton, D. (1997) 'The current state of patent policy in the UK and internationally', *Science, Technology and Innovation*, 10 (1): 17-24.

Nietzsche, F. (1956) *The Birth of Tragedy and The Genealogy of Morals*, Translated by F. Golffing, New York: Doubleday.

Newton, D. (1997) 'The current state of patent policy in the UK and internationally', *Science, Technology and Innovation*, 10 (1): 17-24.

Nogues, J. (1990) 'Patents and Pharmaceutical Drugs: Understanding the Pressures on Developing Countries', *Journal of World Trade*, 24 (6): 81-104.

Noiville, C. (1994) 'Intellectual property rights and life: a comparison of trends in the USA and Europe', Paper presented to International Symposium *Patents, Genes, Butterflies*, Bern: WWF & Swissaid, 20-21 October 1994.

Offe, C. (1985) 'New Social Movements: Challenging the Boundaries of Institutional Politics' *Social Research*, 52 (4).

Pearce, F. (1993), 'Pesticide patent angers Indian Farmers', *New Scientist*, 9/10/93: 7.

Perlman, S. (1991) 'Hegemony and *Arkhe* in Greece: Fourth Century BC Views', in R. Lebow & B. Strauss (ed.), *Hegemonic Rivalry: From Thucydides to the Nuclear Age*, Boulder Co: Westview.

Pimbert, M. (1994), 'The Need for Another Research Paradigm', Seedling, 11 (2): 20-25.

Porter, G. (1992) 'The United States and the Biodiversity Convention: The Case for Participation' Washington: Environment and Energy Study Institute.

Poulantzas, N. (1973), *Political Power and Social Classes*, London: NLB.

Princen, T, and Finger, M. (1994) *Environmental NGOs in World Politics: Linking the Local and the Global*, London: Routledge.

Purdue, D. (1993) *The Trips Story: Intellectual Property, Hegemony and World Trade*, MSc Dissertation, UWE.

Purdue, D. (1995a) 'Hegemonic Trips: World Trade, Intellectual Property and Biodiversity', *Environmental Politics*, 1995, 4 (1): 88-107.

Purdue, D. (1995b) *Globalising Greens: seeds and social movements*, SEPEG Working Paper No. 2, University of the West of England, Bristol.

Purdue, D. (1996) 'Contested Expertise: Social Movements and Plant Biotechnology', *Science as Culture*, 526-545.

Purdue, D. (1998) *Seeds of Change: Conserving Biodiversity and Social Movements*, University of the West of England: Unpublished PhD Thesis.

Purdue, D. (1999) 'Experiments in the governance of biotechnology: a case study of the UK National Consensus Conference', *New Genetics and Society*, 18 (1): 79-99.

Purdue, D. (2000) 'Backyard Biodiversity: Seed Tribes in the West of England', *Science as Culture*, Forthcoming.

Raghavan, C. (1990), *Recolonization, GATT, the Uruguay Round and the Third World*, London and Penang: Zed Books and Third World Network.

Rawls, J. (1973) A *Theory of Justice*, London: O.U. Press.

Reid, W, et al. (1993) *Biodiversity Prospecting: Using Genetic Resources for Sustainable Development*, Washington: World Resources Institute.

Richardson, J. (1986), 'The New Political Economy of Trade' in P. Krugman (ed.), *Strategic Trade Policy and the New Economics*, Cambridge MA and London: MIT Press.

Roberts, T, (1995) 'Draft directive on Legal Protection of Biotechnical Inventions', *European Intellectual Property Review*, 17 (4) April: D-116-7.

Rorty, R. (1989) *Contingency, Irony and Solidarity*, Cambridge: C.U.P.

Rose, C. (1994) 'Greenpeace and the New Environmental Politics' Paper presented at the *Politics of Cultural Change Conference*, Lancaster University, 8-10 July 1994.

Rose, H. (1991) 'Case Studies', in G. Allen and C. Skinner (ed.), *Handbook for Research Students in the Social Sciences*, London: Falmer Press.

Rosenau, J. (1990) *Turbulence in World Politics: A Theory of Change and Continuity*, Hemel Hempstead: Harvester Wheatsheaf.

Schumaker, E.F. (1974) *Small is Beautiful: A Study of Economics as if People Mattered*, London: Abacus.

Scott, J. (1990) *Domination and the Arts of Resistance: Hidden Transcripts*, New Haven and London: Yale U.P.

Scott, J. (1991) *Social Network Analysis: A Handbook*, London: Sage.

Shaw, M. (1993) 'Global Society and Global Responsibility', *Millennium*, 21 (3): 421-34.

Shaw, M. (1994) 'Civil Society and Global Politics: Beyond a Social Movements Approach', *Millennium*, 23 (3): 647-67.

Shiva, V. (1991) *The Violence of the Green Revolution: Third World Agriculture, Ecology and Politics*, London: Zed.

Shiva, V. (1993) *Monocultures of the Mind: Perspectives on Biodiversity and Biotechnology*, London and Penang: Zed Books and Third World Network.

Shiva, V. (1994) 'Local People: Empowerment through control of resources', Paper presented to International Symposium *Patents, Genes, Butterflies*, Bern: WWF & Swissaid, 20-21 October 1994.

Shiva, V, and Holla-Bhar (1993) 'Intellectual Piracy and the Neem Tree', *The Ecologist*, 23 (6): 223-7.

Simmel, G. (1955) 'The Web of Group-Affiliations' in *Conflict and the Web of Group-Affiliations*, New York: The Free Press.

Simmel, G. (1971) 'Sociability' in *On Individuality and Social Forms*, London and Chicago: Chicago University Press.

Simmel, G. (1986) 'Domination and Freedom', in S. Lukes (ed.) *Power*, New York: N.Y.U.P.

Snow, D, and Benford, R. (1988) 'Ideology, Frame Resonance, and Participant Mobilization,' *International Social Movement Research*, 1: 197-217.

Splice of Life. (1995) *Splice of Life* 1, (7): 1-4.

Stoett, P. (1993) 'International Politics and the Protection of Great Whales', *Environmental Politics*, 2 (2): 277-303.

Stoker, G. (1998) 'Governance as theory: five propositions', *International Social Science Journal*, 155: 17-28.

Subramanian, A. (1990) 'TRIPs and the Paradigm of the GATT: a Tropical, Temperate View', *World Economy*, 13 (4): 509-521.

Szerszynski, B. (1997) 'The Varieties of Ecological Piety', *Worldviews: Environment, Culture, Religion*, 1 (1).

Tarrow, S. (1994) *Power in Movement: Social Movements, Collective Action and Politics*, Cambridge: CUP.

Third World Network (1999) 'Joint NGO Statement of Support for the Africa Group Proposals on Reviewing the WTO TRIPs Agreement (Article 27.3)', <http://www.sustain.org/biotech.

Thompson, E.P. (1991) 'Custom, Law and Common Right' in *Customs in Common*, London: Merlin Press.

Thrift, N. (1999) 'Cities and economic change: global governance?' in J. Allen, D. Massey and M. Pryke (ed.), *Unsettling Cities: Movement / Settlement*, London and New York: Routledge and Open University.

Thurston, J. (1993) 'Recent EC Developments in Biotechnology', *European Intellectual Property Rights*, 6: 187-190.

Touraine, A. (1981) *The Voice and the Eye: An analysis of social movements*, Cambridge: C.U.P.

Touraine, A. (1992) 'Beyond Social Movements?', *Theory, Culture and Society*, 9: 125-45.

Touraine, A. (1995) *Critique of Modernity*, Oxford: Blackwell.

United Nations Commission on the Environment and Development. (1992) *Convention on Conservation of Biological Diversity* in Shiva, V. (1993) *Monocultures of the Mind: Perspectives on Biodiversity and Biotechnology*, London and Penang: Zed Books and Third World Network.

Urry, J. (1981) *The Anatomy of Capitalist Societies*, London: MacMillan.

Useem, B, and Zald, M. (1987) 'From Pressure Group to Social Movement: Efforts to Promote Use of Nuclear Power', in M. Zald & J. McCarthy (ed.), *Social Movements*

in an Organizational Society: Collected Essays, New Brunswick & Oxford: Transaction Books.
van Wyk, J. (1995) 'Plant breeders' rights create winners and losers', *Biotechnology and Development Monitor*, 23, June: 15-19.
Vellve, R. (1992) *Saving the Seed: Genetic Diversity and European Agriculture*, London: Earthscan.
Versfeld, M. (1991) *Food for Thought: A Philosopher's Cookbook*, Cape Town: Carrefour Press.
Vidal, J. (1999a) 'Business elite shun Seattle's glare', *The Guardian*, 1/12/1999: 15.
Vidal, J. (1999b) 'Festival of Ideas outside WTO', *The Guardian*, 4/12/1999.
Vidal, J, and Elliot, L. (1999) 'Africa bars the way in Seattle', *The Guardian*, 4/12/1999: 2.
von Molkte, K. (1994) Paper on 'An Agenda for the GATT Trade and Environment Committee' in *Greening World Trade: An Environmental Reform Programme for GATT Conference*, FIELD, University of London, 13 May 1994.
Washburne, N. (1997) 'Oligarchy, Power and New Hybrid Organizing in a Social Movement: a case study of Friends of the Earth', Paper presented to the *BSA Annual Conference*, York University, 7-10 April, 1997.
Watkins, K. (1992) *Fixing the Rules*, London: CIIR.
Weber, M. (1948) 'Politics as a Vocation' in H.H. Gerth and C.Wright Mills (Trans. and Ed) *From Max Weber: Essays in Sociology*, London: Routledge and Keagan Paul.
Weber, M. (1978) 'Charisma and its Transformation', *Economy and Society*, Berkeley: University of California Press.
Welsh, I. (1988) *British Nuclear Power: Protest and Legitimation 1945-1980*, Lancaster University: Unpublished PhD Thesis.
Welsh, I, and McLeish, P. (1996) 'The European Road to Nowhere: Anarchism and Direct Action Against the UK Roads Programme', *Anarchist Studies*, 4(1): 27-44.
Wheale, P, and McNally, R. (1988) *Genetic Engineering: Catastrophe or Utopia?*, London & New York: Harvester Wheatsheaf and St Martin's Press.
Wheale, P, and McNally, R. (1993) 'Biotechnology policy in Europe: a critical evaluation', *Science and Public Policy*, 20 (4): 261-279.
Wilson, E. O, and Peter, F. (ed.) (1988) *Biodiversity*, Washington: National Academy Press.
Wittgenstein, L. (1967) *Philosophical Investigations*, Oxford: Blackwell.
World Bank. (1991) *World Development Report: The Challenge of Development*, Oxford: O.U.P.
Yamin, F. (1993) *Intellectual Property Rights and the Environment: the Role of Patents in the Conservation of Biodiversity*, Unpublished LLM Dissertation, Kings College London.
Yearley, S. (1994) 'Social Movements and Environmental Change', in M. Redclift and T. Benton (ed.), *Social Theory and the Global Environment*, pp 150-168.
Young, O. (1989) 'The politics of international regime formation: managing natural resources and the environment' *International Organization*, 43 (3): 349-76.
Yoxen, E. (1983) *The Gene Business: Who should control Biotechnology?*, London: Pan Books.

Zerner, C. (1994) 'Equity Issues in Bioprospecting', Paper presented to International Symposium *Patents, Genes, Butterflies*, Bern: WWF and Swissaid, 20-21 October 1994.

Index